Weakly Stationary Random Fields, Invariant Subspaces and Applications

Weakly Stationary Random Fields, Invariant Subspaces and Applications

Vidyadhar S. Mandrekar

Michigan State University

East Lansing

David A. Redett

CRC Press

Taylor & Francis Group

Boca Raton   London   New York

CRC Press is an imprint of the
Taylor & Francis Group, an **informa** business

A CHAPMAN & HALL BOOK

CRC Press
Taylor & Francis Group
6000 Broken Sound Parkway NW, Suite 300
Boca Raton, FL 33487-2742

First issued in paperback 2020

© 2018 by Taylor & Francis Group, LLC
CRC Press is an imprint of Taylor & Francis Group, an Informa business

No claim to original U.S. Government works

Version Date: 20171004

ISBN-13: 978-0-367-57272-3 (pbk)
ISBN-13: 978-1-138-56224-0 (hbk)

Library of Congress Cataloging-in-Publication Data

Names: Mandrekar, V. (Vidyadhar), 1939- author. | Redett, David A., author.
Title: Weakly stationary random fields, invariant subspaces and applications
/ Vidyadhar S. Mandrekar, David A. Redett.
Description: Boca Raton : CRC Press, 2018. | Includes bibliographical
references.
Identifiers: LCCN 2017035473 | ISBN 9781138562240 (hardback)
Subjects: LCSH: Random fields. | Invariant subspaces.
Classification: LCC QA274.45 .M3545 2018 | DDC 519.2/32--dc23
LC record available at https://lccn.loc.gov/2017035473

Visit the Taylor & Francis Web site at
http://www.taylorandfrancis.com

and the CRC Press Web site at
http://www.crcpress.com

V. Mandrekar dedicates this book
to the memory of
Professor Pesi Masani
his mathematics godfather.

D. Redett dedicates this book
to his family.

Contents

Introduction

The study of weakly stationary processes was undertaken by H. Wold in [45]. He showed that every weakly stationary process can be decomposed as the sum of two orthogonal weakly stationary processes: one has zero innovation (singular) and the other has nonzero innovation (regular) at every time. Later, Kolmogorov [19] in his seminal paper gave the definition of spectral measure using the Herglotz Theorem (see [12]) and showed that the space generated by a weakly stationary process is isomorphic to the space of square integrable functions with respect to the spectral measure. As a consequence, he obtained analytic conditions for a process to be regular or singular. In [11], Henry Helson presented connections of Kolmogorov's results to invariant subspaces motivated from the work of Helson and Lowdenslager [10], where such a connection is given in the study of multivariate weakly stationary processes. The latter were studied in earlier work of Wiener and Masani ([43], [44]) using the approach of Wold. In his interesting paper [26], Masani observed that both of these approaches are special cases of the Halmos Decomposition of a Hilbert space with respect to an isometry [8]. Here, the Hilbert space can be decomposed in orthogonal subspaces such that on one part the isometry is a pure shift, and on the other part it is unitary. One can find the detailed explanation of this in the works of ([23], [24]). For the convenience of the reader, this is presented in Chapter 1.

In Chapter 1, we present the ideas of Wold and Kolmogorov. To make these ideas more accessible to a larger audience, we present examples of stationary processes and show how they relate to their spectral measures. We postpone the presentation of the Halmos Decomposition and the relation of Kolmogorov's work to invariant subspaces to Chapter 3 as it requires more abstract mathematical ideas. This approach helps to clarify similar problems that come for random fields.

The purpose of this manuscript is to study the extension of this approach to weakly stationary random fields indexed by $\mathbb{Z}^2$. A natural question that one encounters is how to define the "past" and "future" in this case, so that one can define "innovation." This was studied by Helson and Lowdenslager, in [10], where $\mathbb{Z}^2$ was given an ordering through a semigroup. However, as stated in their paper, A. Zygmund asked if one can study in a similar way analytic functions defined with quarter-plane representation. We study here the problems that arise in answering in the positive the question of Zygmund. The approach used here originated in the work of Kallianpur and Mandrekar [18]. In the work of Słociński [39], the Halmos decomposition for a pair of isometries was considered independently. We study here the Halmos Theorem, Wold decomposition for weakly stationary random fields considered in [18] and their applications in analysis for functions defined on the bi-disc. As a consequence we derive the results of Helson and Lowdenslager stated above.

We now describe the organization of the material by chapter. As stated earlier, we present the theory of weakly stationary processes indexed by $\mathbb{Z}$ in Chapter 1 and derive the results of Wold and Kolmogorov, and various results in analysis following the approach in ([23], [24]). The relevant references are given there for the convenience of the reader.

The material in Chapter 2, begins as in Chapter 1 with half-space ordering observing that a weakly stationary random field indexed by $\mathbb{Z}^2$ can be related to two commutative groups of unitary operators indexed by one variable. Following Kallianpur and Mandrekar [18], we define horizontal (vertical) regular and singular processes and their representations. These are used with associated spectral conditions to obtain a fourfold Wold decomposition in Section 2.8. In Section 2.9, under additional assumptions, we present a quarter-plane decomposition connecting with the aforementioned question of Zygmund. In Section 2.10, we derive the results of Helson and Lowdenslager.

In Chapter 3, we begin by presenting the Halmos Decomposition. We use this decomposition to study shift invariant subspaces on $L^2(\mathbb{T})$. Beurling's Theorem regarding the shift invariant subspaces on $H^2(\mathbb{T})$ follows from this work. We then give a fourfold Halmos decomposition for commuting isometries on a Hilbert space. This was proved by Kallianpur and Mandrekar in [18] and Slociński in [39]. Using the ideas as in Helson's book [11] in one variable, we study the problem of invariant subspaces of $L^2(\mathbb{T}^2)$ with respect to normalized Lebesgue measure on the torus deriving the analogue of Beurling's Theorem in $H^2(\mathbb{T}^2)$ defined with quarter-plane coefficients following [6] and relating it to the work of [1].

In Chapter 4, we present some applications to statistical problems. We begin by studying a texture identification problem as a model of random fields with observation noise given by stationary random fields. This problem arises in pattern recognition and diagnostic problems using as instrument. The work was done by Zhang and Mandrekar [46]. It involves the use of Limit theorems in statistics. The other application is to study harmonizable stable random fields which come as models from data transfer load on a server. We use ideas of Cambanis and Miamee [3] and generalizations of Makagon and Mandrekar [22]. The problem of invariant subspaces of $H^p(\mathbb{T}^2)$ arises as a technique. We present work of Redett [33] on this subject with applications to harmonic analysis deriving invariant subspaces of $H^p(\mathbb{T}^2)$ and $L^p(\mathbb{T}^2)$ Also we use moving average representations to give alternative approach to the work Izuchi and Matsugu [16] which answers a question of Rudin [35] on outer functions. The ideas are from [5] which gives more general results.

1

Weakly Stationary Sequences

1.1 Preliminaries

Let Ω be a nonempty set, $\mathcal{F}$ a σ-algebra of subsets of Ω, and P a probability measure defined on $\mathcal{F}$. We will call the triple $(\Omega, \mathcal{F}, P)$ a probability space. All complex-valued, $\mathcal{F}$-measurable functions X on Ω will be called random variables. For a random variable X, we will write EX to denote the integral $\int_\Omega X\,dP$. A sequence, X_n, $n \in \mathbb{Z}$, of random variable will be called a random sequence. We say a random sequence X_n, $n \in \mathbb{Z}$ is a second order sequence, if $E|X_n|^2 < \infty$, $n \in \mathbb{Z}$. Recall that such a sequence is contained in $L^2(\Omega, \mathcal{F}, P)$. A second order sequence X_n, $n \in \mathbb{Z}$ is called a weakly stationary sequence, if

1. $EX_n = C$, $n \in \mathbb{Z}$, and

2. $\text{cov}(X_n, X_m) := E(X_n - EX_n)\overline{(X_m - EX_m)}$, $n, m \in \mathbb{Z}$ depends only on $n - m$. That is,

$$\text{cov}(X_{n+k}, X_k) = \text{cov}(X_n, X_0), \qquad n, k \in \mathbb{Z}. \tag{1.1}$$

For the sake of simplicity, we assume that $EX_n = 0$, $n \in \mathbb{Z}$. We point out that, under this assumption, the $\text{cov}(X_n, X_m)$ is just the inner product of X_n and X_m in $L^2(\Omega, \mathcal{F}, P)$. For a weakly stationary sequence X_n, $n \in \mathbb{Z}$, we define the covariance function for the sequence to be

$$r(n) = \text{cov}(X_n, X_0), \qquad n \in \mathbb{Z}. \tag{1.2}$$

We close this section by explaining why the random sequences in this section are called "weakly" stationary, rather than just stationary. The name suggests that there must be a stronger type of stationarity. This is indeed the case. A random sequence X_n, $n \in \mathbb{Z}$ is called a stationary sequence, if for any finite collection $n_1, ..., n_k \in \mathbb{Z}$ and any $l \in \mathbb{Z}$ the finite-dimensional distributions of $X_{n_1}, ... X_{n_k}$ and $X_{n_1+l}, ... X_{n_k+l}$ coincide. It follows from this definition that if X_n, $n \in \mathbb{Z}$ is a stationary sequence and a second order sequence, then X_n, $n \in \mathbb{Z}$ is a weakly stationary sequence. Just so we have at least one example of a stationary sequence, note that every sequence of i.i.d. random variables is a stationary sequence. We also point out that for complex Gaussian sequences, weakly stationarity is equivalent to stationarity.

1.2 Examples of Weakly Stationary Sequences

We now turn to some examples. In the following examples, we will use L_0^2 to denote the set of all $\zeta \in L^2(\Omega, \mathcal{F}, P)$ with the property that $E\zeta = 0$.

Example 1.2.1. *Let $\zeta \in L_0^2$, $\gamma \in [-\pi, \pi)$ and define $X_n = \zeta e^{in\gamma}$, $n \in \mathbb{Z}$. It is straightforward to check that X_n is a weakly stationary sequence with covariance function $r(n) = e^{in\gamma} E|\zeta|^2$, $n \in \mathbb{Z}$.*

Example 1.2.2. *Let $Z_1, ..., Z_N$ be an orthogonal collection of random variables in L_0^2. That is, $\text{cov}(Z_l, Z_k) = 0$ for all $k \neq l$. Now, let $\gamma_1, ..., \gamma_N \in [-\pi, \pi)$ and define $X_n = \sum_{k=1}^{N} Z_k e^{in\gamma_k}$, $n \in \mathbb{Z}$. X_n is a weakly stationary sequence with covariance function $r(n) = \sum_{k=1}^{N} e^{in\gamma_k} E|Z_k|^2$, $n \in \mathbb{Z}$.*

Example 1.2.3 (White Noise). *Take X_n, $n \in \mathbb{Z}$ to be an orthonormal sequence of random variables in L_0^2. That is, $\text{cov}(X_n, X_m) = \delta_{nm} = \begin{cases} 1, & n = m \\ 0, & n \neq m \end{cases}$. This sequence is often referred to as white noise. This sequence is weakly stationary with covariance function $r(n) = \delta_{n0}$, $n \in \mathbb{Z}$.*

Example 1.2.4 (Moving Average Sequence). *Let $a_k \in \mathbb{C}$, $k \in \mathbb{Z}$ with $\sum_{k=-\infty}^{\infty} |a_k|^2 < \infty$ and let ξ_n, $n \in \mathbb{Z}$ be white noise. Take $X_n = \sum_{k=-\infty}^{\infty} a_k \xi_{n-k}$, $n \in \mathbb{Z}$. This sequence is called a moving average sequence. This sequence is weakly stationary with covariance function $r(n) = \sum_{k=-\infty}^{\infty} a_k \bar{a}_{k-n} = \sum_{k=-\infty}^{\infty} a_{n+k} \bar{a}_k$, $n \in \mathbb{Z}$. Note: If $a_k = 0$ for $k = -1, -2, ...$, then $X_n = \sum_{k=0}^{\infty} a_k \xi_{n-k}$, $n \in \mathbb{Z}$ is called a one-sided moving average sequence.*

1.3 Spectral Representation of the Covariance Function

We begin this section by observing some properties of the covariance function for a weakly stationary sequence. Let X_n, $n \in \mathbb{Z}$ be a weakly stationary sequence with covariance function $r(n)$, $n \in \mathbb{Z}$. First, we observe the following.

1. $\underline{r(-n)} = \text{cov}(X_{-n}, X_0) = \text{cov}(X_0, X_n) = EX_0\overline{X_n} = \overline{EX_n\overline{X_0}} = \underline{r(n)}$

2. $r(0) = \text{cov}(X_0, X_0) = E|X_0|^2 \geq 0$

3. For every finite collection $t_1, ..., t_m \in \mathbb{Z}$ and complex numbers $a_1, ..., a_m$, $\displaystyle\sum_{k=1}^{m}\sum_{j=1}^{m} a_k \overline{a}_j r(t_k - t_j) = \sum_{k=1}^{m}\sum_{j=1}^{m} a_k \overline{a}_j E X_{t_k} \overline{X}_{t_j} =$

$$E \left| \sum_{j=1}^{m} a_j X_{t_j} \right|^2 \geq 0.$$

Property 3 states that r is a positive definite function. As such, we may employ a theorem due to G. Herglotz (see [12]), which states that there exists a unique (finite) positive measure F_X defined on $\mathcal{B}(\mathbb{T})$, the Borel σ-algebra of $\mathbb{T} = \left\{ e^{i\lambda} : \lambda \in [-\pi, \pi) \right\}$, such that

$$r(n) = \int_{[-\pi, \pi)} e^{in\lambda} dF_X(e^{i\lambda}). \tag{1.3}$$

This measure F_X is called the spectral measure of the weakly stationary sequence X_n, $n \in \mathbb{Z}$.

1.4 Examples of Spectral Measures

We now revisit our examples of weakly stationary sequences in the previous section and identify their spectral measures. Recall that we used L_0^2 to denote the set of all $\zeta \in L^2(\Omega, \mathcal{F}, P)$ with the property that $E\zeta = 0$.

Example 1.4.1. *Let $\zeta \in L_0^2$, $\gamma \in [-\pi, \pi)$ and define $X_n = \zeta e^{in\gamma}$, $n \in \mathbb{Z}$. We already noted that X_n has covariance function $r(n) = e^{in\gamma} E|\zeta|^2$, $n \in \mathbb{Z}$. We now note that X_n has spectral measure F_X that is a discrete measure concentrated on the point $e^{i\gamma}$ with mass $F_X(\{e^{i\gamma}\}) = E|\zeta|^2$.*

Example 1.4.2. *Let $Z_1, ..., Z_N$ be an orthogonal collection of random variables in L_0^2, $\gamma_1, ..., \gamma_N \in [-\pi, \pi)$ and define $X_n = \displaystyle\sum_{k=1}^{N} Z_k e^{in\gamma_k}$, $n \in \mathbb{Z}$. We already saw that X_n has covariance function $r(n) = \displaystyle\sum_{k=1}^{N} e^{in\gamma_k} E|Z_k|^2$, $n \in \mathbb{Z}$. We now point out that it has spectral measure F_X that is a discrete measure concentrated at the points $e^{i\gamma_1}, ..., e^{i\gamma_N}$ with masses $F_X(\{e^{\gamma_k}\}) = E|Z_k|^2$.*

Example 1.4.3 (White Noise). *In this example and henceforth, we will use σ to denote Lebesgue measure on $\mathbb{T}$, normalized so that $\sigma(\mathbb{T}) = 1$.*

Take X_n, $n \in \mathbb{Z}$ to be an orthonormal sequence of random variables in L_0^2. This sequence has covariance function $r(n) = \delta_{n0}$, $n \in \mathbb{Z}$. The spectral measure for this sequence is $F_X = \sigma$.

Example 1.4.4 (Moving Average Sequence). *Let* $a_k \in \mathbb{C}$, $k \in \mathbb{Z}$ *with* $\sum_{k=-\infty}^{\infty} |a_k|^2 < \infty$ *and let* ξ_n, $n \in \mathbb{Z}$ *be white noise. Take* $X_n = \sum_{k=-\infty}^{\infty} a_k \xi_{n-k}$, $n \in \mathbb{Z}$. *This process has covariance function* $r(n) = \sum_{k=-\infty}^{\infty} a_k \bar{a}_{k-n}$, $n \in \mathbb{Z}$. *The spectral measure for this sequence is the measure* F_X, *which is absolutely continuous with respect to* σ, *with density function* $f_X(e^{i\lambda}) = \left| \sum_{k=-\infty}^{\infty} \bar{a}_k e^{ik\lambda} \right|^2$.
If we are only considering a one-sided moving average sequence, $X_n = \sum_{k=0}^{\infty} a_k \xi_{n-k}$, $n \in \mathbb{Z}$, *then we get spectral measure* F_X, *which is once again absolutely continuous with respect to* σ, *with density* $f_X(e^{i\lambda}) = \left| \sum_{k=0}^{\infty} \bar{a}_k e^{ik\lambda} \right|^2$.
We will look at this example more closely in a later section.

1.5 Canonical Isomorphism between $L^2(\mathbb{T}, F_X)$ and $L(X)$

The isomorphism established in this section is a major step in the direction of the spectral representation of stationary sequences. However, it is of interest on its own. It shows how one can transform a prediction problem from $L(X)$ to the function space $L^2(\mathbb{T}, F_X)$ and then back.

Given a weakly stationary sequence X_n, $n \in \mathbb{Z}$, let $L(X)$ denote the (closed) subspace of $L^2(\Omega, \mathcal{F}, P)$ generated by X_n, $n \in \mathbb{Z}$. That is,

$$L(X) = \overline{\operatorname{span}}\{X_n : n \in \mathbb{Z}\}.$$

Let F_X be the spectral measure for this process. We will write $L^2(\mathbb{T}, F_X)$ to denote the collection of all measurable complex-valued functions defined on $\mathbb{T}$ and square integrable with respect to F_X. As always, we will identify functions that are equal $[F_X]$-a.e.

Since $L(X)$ and $L^2(\mathbb{T}, F_X)$ are separable Hilbert spaces, there are many isomorphisms between them. We are interested in one particular isomorphism. Define $\mathcal{J} : L^2(\mathbb{T}, F_X) \to L(X)$ by $\mathcal{J}(e^{in\lambda}) = X_n$. Extending this by linearity, we have for a finite linear combination that

$$\mathcal{J}\left(\sum_k a_k e^{ik\lambda}\right) = \sum_k a_k X_k.$$

We now have that $\mathcal{J}$ is a linear mapping between the collection of all finite combinations of $\{e^{in\lambda} : n \in \mathbb{Z}\}$ and the collection of all finite combinations of

$\{X_n : n \in \mathbb{Z}\}$. This mapping is clearly onto. We also see that $\mathcal{J}$ preserves the inner product since,

$$
\begin{aligned}
\left(\mathcal{J}\left(\sum_k a_k e^{ik\lambda}\right), \mathcal{J}\left(\sum_l b_l e^{il\lambda}\right)\right)_{L(X)} &= \left(\sum_k a_k X_k, \sum_l b_l X_l\right)_{L(X)} \\
&= \sum_k \sum_l a_k \bar{b}_l (X_k, X_l)_{L(X)} \\
&= \sum_k \sum_l a_k \bar{b}_l r(k-l) \\
&= \sum_k \sum_l a_k \bar{b}_l \int_{[-\pi,\pi)} e^{i(k-l)\lambda} dF_X(e^{i\lambda}) \\
&= \int_{[-\pi,\pi)} \sum_k a_k e^{ik\lambda} \overline{\sum_l b_l e^{il\lambda}} dF_X(e^{i\lambda}) \\
&= \left(\sum_k a_k e^{ik\lambda}, \sum_l b_l e^{il\lambda}\right)_{L^2(\mathbb{T},F_X)}
\end{aligned}
$$

and is therefore a one-to-one mapping. We now have that $\mathcal{J}$ is an isomorphism between two dense linear manifolds of $L^2(\mathbb{T}, F_X)$ and $L(X)$, respectively, and as such may be extended to an isomorphism, which we will still call $\mathcal{J}$, from $L^2(\mathbb{T}, F_X)$ to $L(X)$. We call this isomorphism our canonical isomorphism.

1.6 Spectral Representation of a Weakly Stationary Sequence

In this section, we will take a closer look at our canonical isomorphism from the last section. We will see that it can be represented by a stochastic integral. We will use this to give the spectral representation of a stationary sequence. The appendix contains a review of orthogonally scattered set functions and stochastic integration.

Recall that $\mathcal{B}(\mathbb{T})$ denotes the Borel σ-algebra of $\mathbb{T}$. Now, define $\mathcal{Z}_X : \mathcal{B}(\mathbb{T}) \to L(X)$ by

$$
\mathcal{Z}_X(\Delta) = \mathcal{J}(\chi_\Delta),
$$

where $\mathcal{J}$ denotes our canonical isomorphism from $L^2(\mathbb{T}, F_X)$ to $L(X)$. $\mathcal{Z}_X$ is an orthogonally scattered set function. Indeed, if Δ_1 and Δ_2 are disjoint sets in $\mathcal{B}(\mathbb{T})$, then

$$
\begin{aligned}
(\mathcal{Z}_X(\Delta_1), \mathcal{Z}_X(\Delta_2))_{L(X)} &= (\mathcal{J}(\chi_{\Delta_1}), \mathcal{J}(\chi_{\Delta_2}))_{L(X)} \\
&= (\chi_{\Delta_1}, \chi_{\Delta_2})_{L^2(\mathbb{T},F_X)} \\
&= \int_{[-\pi,\pi)} \chi_{\Delta_1} \chi_{\Delta_2} \, dF_X(e^{i\lambda}) = 0,
\end{aligned}
$$

and if $(\Delta_n)_{n=1}^\infty$ are disjoint sets in $\mathcal{B}(\mathbb{T})$, then

$$\mathcal{Z}_X\left(\cup_{n=1}^\infty \Delta_n\right) = \mathcal{J}\left(\chi_{\cup_{n=1}^\infty \Delta_n}\right) = \mathcal{J}\left(\sum_{n=1}^\infty \chi_{\Delta_n}\right) = \sum_{n=1}^\infty \mathcal{J}\left(\chi_{\Delta_n}\right) = \sum_{n=1}^\infty \mathcal{Z}_X(\Delta_n).$$

We also point out that for $\Delta \in \mathcal{B}(\mathbb{T})$,

$$\|\mathcal{Z}_X(\Delta)\|^2 = \int_{[-\pi,\pi)} \chi_\Delta^2 \, dF_X\left(e^{i\lambda}\right) = F_X(\Delta).$$

Therefore, F_X is the control measure for $\mathcal{Z}_X$. Now that this has been established, we leave it to the reader to show that for every $f \in L^2(\mathbb{T}, F_X)$,

$$\mathcal{J}(f) = \int_{[-\pi,\pi)} f\left(e^{i\lambda}\right) d\mathcal{Z}_X\left(e^{i\lambda}\right).$$

We suggest that the reader start with simple functions, then use the fact that simple functions are dense in $L^2(\mathbb{T}, F_X)$ to get the complete result. We point out that this result implies the spectral representation theorem for weakly stationary sequences which states that if X_n, $n \in \mathbb{Z}$ is a weakly stationary sequence, then

$$X_n = \int_{[-\pi,\pi)} e^{in\lambda} \, d\mathcal{Z}_X\left(e^{i\lambda}\right),$$

for all $n \in \mathbb{Z}$, where $\mathcal{Z}_X$ is an orthogonally scattered set function and this representation is unique.

1.7   The Shift Operator on $L(X)$

In this section, we will show the existence of a unique unitary operator on $L(X)$ that takes $X_n \mapsto X_{n+1}$ for a weakly stationary sequence X_n, $n \in \mathbb{Z}$. We will then use this operator and the spectral theorem for unitary operators to give an alternative approach to recovering the spectral representation theorem for weakly stationary sequences.

Let X_n, $n \in \mathbb{Z}$ be a weakly stationary sequence and let $L(X)$ be as defined above. For each $n \in \mathbb{Z}$, define $UX_n = X_{n+1}$. Extending by linearity, we get for a finite linear combination that

$$U\left(\sum_k a_k X_k\right) = \sum_k a_k X_{k+1}.$$

To see that U is well defined, note that if

$$\sum_k a_k X_k = \sum_k b_k X_k$$

then

$$\left\| U\left(\sum_k a_k X_k\right) - U\left(\sum_k b_k X_k\right) \right\|_{L(X)}^2$$

$$= \left\| \sum_k a_k X_{k+1} - \sum_k b_k X_{k+1} \right\|_{L(X)}^2$$

$$= \sum_k \sum_l (a_k - b_k)\overline{(a_l - b_l)}\, (X_{k+1}, X_{l+1})_{L(X)}$$

$$= \sum_k \sum_l (a_k - b_k)\overline{(a_l - b_l)}\, (X_k, X_l)_{L(X)}$$

$$= \left\| \sum_k a_k X_k - \sum_k b_k X_k \right\|_{L(X)}^2 = 0.$$

Note that the third equality follows from stationarity. Therefore, U is well defined on the collection of all finite linear combinations of X_n, $n \in \mathbb{Z}$. We also see that U preserves the inner product.

$$\left(U\left(\sum_k a_k X_k\right), U\left(\sum_k b_k X_k\right) \right)_{L(X)} = \left(\sum_k a_k X_{k+1}, \sum_k b_k X_{k+1} \right)_{L(X)}$$

$$= \sum_k \sum_l a_k \bar{b}_k\, (X_{k+1}, X_{l+1})_{L(X)}$$

$$= \sum_k \sum_l a_k \bar{b}_k\, (X_k, X_l)_{L(X)}$$

$$= \left(\sum_k a_k X_k, \sum_k b_k X_k \right)_{L(X)}$$

Note that the third equality follows from stationarity. Therefore, we have a linear operator U that preserves the inner product, which is defined on a dense linear manifold of $L(X)$. Therefore, as we did above, we can extend U to all of $L(X)$. This extension, which we will also call U, is our desired unitary operator. It is left to show that U is onto. To see this, let $\zeta \in L(X)$. Then, ζ may be written as the following limit.

$$\zeta = \lim_n \sum_k a_k^{(n)} X_k.$$

Using the same type of calculations as above, one can show that $\left(\sum_k a_k^{(n)} X_{k-1} \right)_n$ is a Cauchy sequence in $L(X)$ and hence has a limit. Let us call this limit η. It is straightforward to see that $U\eta = \zeta$. Therefore, U is onto.

We now tie this into the spectral representation of weakly stationary sequences. Given a weakly stationary sequence X_n, $n \in \mathbb{Z}$, our work above guarantees the existence of a unitary operator on $L(X)$ with the property that $UX_n = X_{n+1}$, for all $n \in \mathbb{Z}$. This observation shows that $X_n = U^n X_0$, for all $n \in \mathbb{Z}$. Since U is a unitary operator, we may employ the spectral theorem for unitary operators to write U as

$$U = \int_{[-\pi,\pi)} e^{i\lambda} \, d\mathcal{E}(e^{i\lambda}),$$

where $\mathcal{E}$ is a resolution of the identity. Using the symbolic calculus for normal operators, we get that

$$U^n = \int_{[-\pi,\pi)} e^{in\lambda} \, d\mathcal{E}(e^{i\lambda}).$$

Now, we define $\mathcal{Z}_X : \mathcal{B}(\mathbb{T}) \to L(X)$ by $\mathcal{Z}_X(\Delta) = \mathcal{E}(\Delta)X_0$. It is shown in the appendix that $\mathcal{Z}_X$ is an orthogonally scattered set function. Putting this all together, we get the spectral representation of a weakly stationary sequence. That is,

$$X_n = U^n X_0 = \int_{[-\pi,\pi)} e^{in\lambda} \, d\mathcal{E}(e^{i\lambda})X_0 = \int_{[-\pi,\pi)} e^{in\lambda} \, d\mathcal{Z}_X(e^{i\lambda}).$$

1.8 Moving Averages and Densities

One of the most popular examples of a weakly stationary sequence is the moving average sequence. We first introduced this sequence in Section 1.2. In this section, we give a characterization of moving average sequences in the context of the sequence's spectral measure.

Let $(\Omega, \mathcal{F}, P)$ be a probability space. Let X_n, $n \in \mathbb{Z}$ be a random sequence. Suppose further that X_n, $n \in \mathbb{Z}$ is a moving average sequence; that is, there exists a sequence of complex numbers $(a_n)_{n \in \mathbb{Z}}$ with the property that $\sum_{n \in \mathbb{Z}} |a_n|^2 < \infty$ and a white noise sequence ξ_n, $n \in \mathbb{Z}$ in $L^2(\Omega, \mathcal{F}, P)$ such that

$$X_n = \sum_{k \in \mathbb{Z}} a_k \, \xi_{n-k},$$

in $L^2(\Omega, \mathcal{F}, P)$, for all $n \in \mathbb{Z}$. A straightforward calculation shows that X_n, $n \in \mathbb{Z}$ is a weakly stationary sequence.

We begin by examining the spectral measure for this sequence. First recall that white noise is a stationary sequence and as such has a spectral representation.

$$\xi_n = \int_{[-\pi,\pi)} e^{in\lambda} dZ_\xi(e^{i\lambda}),$$

where Z_ξ is the orthogonally scattered set function associated with the sequence ξ_n, $n \in \mathbb{Z}$. We also showed in an earlier example that F_ξ, the spectral measure for ξ_n, $n \in \mathbb{Z}$, is normalized Lebesgue measure on $\mathbb{T}$. That is, $dF_\xi(e^{i\lambda}) = d\sigma(e^{i\lambda})$. We now calculate the covariance function for X_n, $n \in \mathbb{Z}$.

$$
\begin{aligned}
r(n) &= (X_n, X_0)_{L^2(\Omega, \mathcal{F}, P)} \\
&= \left(\sum_{k \in \mathbb{Z}} a_k \xi_{n-k}, \sum_{k \in \mathbb{Z}} a_k \xi_{-k} \right)_{L^2(\Omega, \mathcal{F}, P)} \\
&= \left(\sum_{k \in \mathbb{Z}} a_k \int_{[-\pi,\pi)} e^{i(n-k)\lambda} dZ_\xi(e^{i\lambda}), \sum_{k \in \mathbb{Z}} a_k \int_{[-\pi,\pi)} e^{-ik\lambda} dZ_\xi(e^{i\lambda}) \right)_{L^2(\Omega, \mathcal{F}, P)} \\
&= \left(\int_{[-\pi,\pi)} \sum_{k \in \mathbb{Z}} a_k e^{i(n-k)\lambda} dZ_\xi(e^{i\lambda}), \int_{[-\pi,\pi)} \sum_{k \in \mathbb{Z}} a_k e^{-ik\lambda} dZ_\xi(e^{i\lambda}) \right)_{L^2(\Omega, \mathcal{F}, P)} \\
&= \int_{[-\pi,\pi)} \sum_{k \in \mathbb{Z}} a_k e^{i(n-k)\lambda} \overline{\sum_{k \in \mathbb{Z}} a_k e^{-ik\lambda}} \, dF_\xi(e^{i\lambda}) \\
&= \int_{[-\pi,\pi)} \left| \sum_{k \in \mathbb{Z}} a_k e^{-ik\lambda} \right|^2 e^{in\lambda} \, d\sigma(e^{i\lambda}).
\end{aligned}
$$

Therefore, we see that the spectral measure of X_n, $n \in \mathbb{Z}$, is absolutely continuous with respect to normalized Lebesgue measure with density $f_X(e^{i\lambda}) = \left| \sum_{k \in \mathbb{Z}} a_k e^{-ik\lambda} \right|^2 = \left| \sum_{k \in \mathbb{Z}} \overline{a_k} e^{ik\lambda} \right|^2$. We now record this observation as a theorem.

Theorem 1.8.1. *Suppose that X_n, $n \in \mathbb{Z}$ is a moving average sequence; that is, there exists a sequence of complex numbers $(a_n)_{n \in \mathbb{Z}}$ with the property that $\sum_{n \in \mathbb{Z}} |a_n|^2 < \infty$ and a white noise sequence ξ_n, $n \in \mathbb{Z}$ in $L^2(\Omega, \mathcal{F}, P)$ such that*

$$X_n = \sum_{k \in \mathbb{Z}} a_k \, \xi_{n-k},$$

in $L^2(\Omega, \mathcal{F}, P)$, for all $n \in \mathbb{Z}$. Then, X_n, $n \in \mathbb{Z}$ is a weakly stationary sequence that has a spectral measure that is absolutely continuous with respect to normalized Lebesgue measure on $\mathbb{T}$ with density $f_X(e^{i\lambda}) = |\varphi(e^{i\lambda})|^2$, where $\varphi \in L^2(\mathbb{T}, \sigma)$ and $\hat{\varphi}(k) = \overline{a_k}$ for all $k \in \mathbb{Z}$.

Now, let us examine the situation from the other direction. Suppose that X_n, $n \in \mathbb{Z}$ is a weakly stationary sequence with spectral measure $dF_X(e^{i\lambda}) = |\varphi(e^{i\lambda})|^2 \, d\sigma(e^{i\lambda})$, where $\varphi \in L^2(\mathbb{T}, \sigma)$. As such, φ is equal to its Fourier series in $L^2(\mathbb{T}, \sigma)$. That is,

$$\varphi(e^{i\lambda}) = \sum_{k \in \mathbb{Z}} \hat{\varphi}(k) e^{ik\lambda} \text{ in } L^2(\mathbb{T}, \sigma).$$

We now define the linear transformation $W : L(X) \to L^2(\mathbb{T}, \sigma)$ by $W(X_n) = e^{in\lambda}\overline{\varphi}(e^{i\lambda})$. W is an isometry and as such, we may identify $L(X)$ and $W(L(X))$, as well as, X_n and $e^{in\lambda}\overline{\varphi}(e^{i\lambda})$. $e^{in\lambda}\overline{\varphi}(e^{i\lambda})$ has a moving average representation since

$$e^{in\lambda}\overline{\varphi}(e^{i\lambda}) = \sum_{k \in \mathbb{Z}} \overline{\hat{\varphi}(k)} e^{i(n-k)\lambda} \text{ in } L^2(\mathbb{T}, \sigma).$$

So, we may conclude the X_n, $n \in \mathbb{Z}$ has a moving average representation of the form

$$X_n = \sum_{k \in \mathbb{Z}} \overline{\hat{\varphi}(k)} \xi_{n-k},$$

for some white noise sequence $(\xi_n)_{n \in \mathbb{Z}}$ contained in some Hilbert space that contains $L(X)$ as a subspace. A natural question at this point might be: when is this white noise sequence contained in $L(X)$? Let us examine that question. Let

$$L(\xi) = \overline{\text{span}}\left\{\xi_n : n \in \mathbb{Z}\right\}.$$

Now, define the linear transformation $W^{\#} : L(\xi) \to L^2(\mathbb{T}, \sigma)$ by $W^{\#}(\xi_n) = e^{in\lambda}$. $W^{\#}$ is clearly an isomorphism. In a similar manner, we define

$$L(X) = \overline{\text{span}}\left\{X_n : n \in \mathbb{Z}\right\}.$$

It follows from the moving average representation of X_n, $n \in \mathbb{Z}$, that $L(X) \subseteq L(\xi)$. Again from this moving average representation, we see that

$$W^{\#}(X_n) = e^{in\lambda}\overline{\varphi}(e^{i\lambda}).$$

It follows then that

$$W^{\#}(L(X)) = \overline{\text{span}}\left\{e^{in\lambda}\overline{\varphi}(e^{i\lambda}) : n \in \mathbb{Z}\right\},$$

which is a doubly invariant subspace of $L^2(\mathbb{T}, \sigma)$. By Theorem 3.2.1, all doubly invariant subspaces of $L^2(\mathbb{T}, \sigma)$ are of the form $\mathbf{1}_E L^2(\mathbb{T}, \sigma)$, where E is a Lebesgue measurable subset of $\mathbb{T}$. In our particular case, $E = \left\{e^{i\lambda} \in \mathbb{T} : \varphi(e^{i\lambda}) \neq 0\right\}$. From this observation, we see that if E^c, the complement of E, has Lebesgue measure zero, then $W^{\#}(L(X)) = L^2(\mathbb{T}, \sigma)$. This together with the fact that $W^{\#}$ is an isomorphism gives us that $L(X) = L(\xi)$ and therefore, $\xi_n \in L(X)$ for all $n \in \mathbb{Z}$. If, on the other hand, the Lebesgue measure of E^c is positive, then $L(X) \subsetneq L(\xi)$. So, there exists an $m \in \mathbb{Z}$ such that $\xi_m \notin L(X)$. An important observation to make here, if not already observed, is that $|\varphi(e^{i\lambda})|^2$ is the density of the spectral measure of our sequence X_n, $n \in \mathbb{Z}$. Therefore, whether or not our spectral density vanishes on a set of positive measure impacts the location of our white noise sequence.

We will now record these observations as a theorem.

Theorem 1.8.2. *Suppose that X_n, $n \in \mathbb{Z}$ is a weakly stationary sequence with spectral measure $dF_X(e^{i\lambda}) = |\varphi(e^{i\lambda})|^2 \, d\sigma(e^{i\lambda})$, where $\varphi \in L^2(\mathbb{T}, \sigma)$. Then, X_n, $n \in \mathbb{Z}$ may be represented as a moving average sequence; that is,*

$$X_n = \sum_{k \in \mathbb{Z}} \overline{\hat{\varphi}(k)} \, \xi_{n-k},$$

in some Hilbert space $\mathcal{H}$ containing the white noise sequence ξ_n, $n \in \mathbb{Z}$ and having $L(X)$ as a subspace. If $\varphi \neq 0$ $[\sigma]$-a.e., then our white noise sequence is contained in $L(X)$.

We now examine the one-sided moving average sequences. Although much of the work that follows is very similar to that of the moving average sequences, we include it for completeness.

Let $(\Omega, \mathcal{F}, P)$ be a probability space. Let X_n, $n \in \mathbb{Z}$ be a random sequence. Suppose further that X_n, $n \in \mathbb{Z}$ is a one-sided moving average sequence. That is, there exists a sequence of complex numbers $(a_n)_{n \in \mathbb{N}}$ with the property that $\sum_{n \in \mathbb{N}} |a_n|^2 < \infty$ and a white noise sequence ξ_n, $n \in \mathbb{Z}$ in $L^2(\Omega, \mathcal{F}, P)$ such that

$$X_n = \sum_{k \in \mathbb{N}} a_k \, \xi_{n-k},$$

in $L^2(\Omega, \mathcal{F}, P)$, for all $n \in \mathbb{Z}$. A straightforward calculation shows that X_n, $n \in \mathbb{Z}$ is a weakly stationary sequence.

We begin by examining the spectral measure for this sequence. First recall that white noise is a stationary sequence and as such has a spectral representation.

$$\xi_n = \int_{[-\pi, \pi)} e^{in\lambda} dZ_\xi(e^{i\lambda}),$$

where Z_ξ is the orthogonally scattered set function associated with the sequence ξ_n, $n \in \mathbb{Z}$. We also showed in an earlier example that F_ξ, the spectral measure for ξ_n, $n \in \mathbb{Z}$, is normalized Lebesgue measure on $\mathbb{T}$. That is,

$dF_\xi(e^{i\lambda}) = d\sigma(e^{i\lambda})$. We now calculate the covariance function for X_n, $n \in \mathbb{Z}$.

$$r(n) = (X_n, X_0)_{L^2(\Omega,\mathcal{F},P)}$$

$$= \left(\sum_{k\in\mathbb{N}} a_k \xi_{n-k}, \sum_{k\in\mathbb{N}} a_k \xi_{-k} \right)_{L^2(\Omega,\mathcal{F},P)}$$

$$= \left(\sum_{k\in\mathbb{N}} a_k \int_{[-\pi,\pi)} e^{i(n-k)\lambda} dZ_\xi(e^{i\lambda}), \sum_{k\in\mathbb{N}} a_k \int_{[-\pi,\pi)} e^{-ik\lambda} dZ_\xi(e^{i\lambda}) \right)_{L^2(\Omega,\mathcal{F},P)}$$

$$= \left(\int_{[-\pi,\pi)} \sum_{k\in\mathbb{N}} a_k e^{i(n-k)\lambda} dZ_\xi(e^{i\lambda}), \int_{[-\pi,\pi)} \sum_{k\in\mathbb{N}} a_k e^{-ik\lambda} dZ_\xi(e^{i\lambda}) \right)_{L^2(\Omega,\mathcal{F},P)}$$

$$= \int_{[-\pi,\pi)} \sum_{k\in\mathbb{N}} a_k e^{i(n-k)\lambda} \overline{\sum_{k\in\mathbb{Z}} a_k e^{-ik\lambda}} \, dF_\xi(e^{i\lambda})$$

$$= \int_{[-\pi,\pi)} \left| \sum_{k\in\mathbb{N}} a_k e^{-ik\lambda} \right|^2 e^{in\lambda} \, d\sigma(e^{i\lambda}).$$

Therefore, we see that the spectral measure of X_n, $n \in \mathbb{Z}$, is absolutely continuous with respect to normalized Lebesgue measure with density $f_X(e^{i\lambda}) =$

$$\left| \sum_{k\in\mathbb{N}} a_k e^{-ik\lambda} \right|^2 = \left| \sum_{k\in\mathbb{N}} \overline{a_k} e^{ik\lambda} \right|^2.$$ We now record this observation as a theorem.

Theorem 1.8.3. *Suppose that X_n, $n \in \mathbb{Z}$ is a one-sided moving average sequence; that is, there exists a sequence of complex numbers $(a_n)_{n\in\mathbb{N}}$ with the property that $\sum_{n\in\mathbb{N}} |a_n|^2 < \infty$ and a white noise sequence ξ_n, $n \in \mathbb{Z}$ in $L^2(\Omega, \mathcal{F}, P)$ such that*

$$X_n = \sum_{k\in\mathbb{N}} a_k \, \xi_{n-k},$$

in $L^2(\Omega, \mathcal{F}, P)$, for all $n \in \mathbb{Z}$. Then, X_n, $n \in \mathbb{Z}$ is a weakly stationary sequence that has a spectral measure that is absolutely continuous with respect to normalized Lebesgue measure with density $f_X(e^{i\lambda}) = |\varphi(e^{i\lambda})|^2$, where $\varphi \in H^2(\mathbb{T}) = \overline{\text{span}}\{e^{in\lambda} : n \geq 0\}$ in $L^2(\mathbb{T}, \sigma)$ and $\hat{\varphi}(k) = \overline{a_k}$ for all $k \in \mathbb{N}$.

Now, let us examine the situation from the other direction. Suppose that X_n, $n \in \mathbb{Z}$ is a weakly stationary sequence with spectral measure $dF_X(e^{i\lambda}) = |\varphi(e^{i\lambda})|^2 \, d\sigma(e^{i\lambda})$, where $\varphi \in H^2(\mathbb{T})$. As such, φ is equal to its Fourier series in $L^2(\mathbb{T}, \sigma)$. That is,

$$\varphi(e^{i\lambda}) = \sum_{k\in\mathbb{N}} \hat{\varphi}(k) e^{ik\lambda} \text{ in } L^2(\mathbb{T}, \sigma).$$

We now define the linear transformation $W : L(X) \to L^2(\mathbb{T}, \sigma)$ by $W(X_n) =$

$e^{in\lambda}\overline{\varphi}(e^{i\lambda})$. W is an isometry and as such, we may identify $L(X)$ and $W(L(X))$, as well as, X_n and $e^{in\lambda}\overline{\varphi}(e^{i\lambda})$. $e^{in\lambda}\overline{\varphi}(e^{i\lambda})$ has a moving average representation since

$$e^{in\lambda}\overline{\varphi}(e^{i\lambda}) = \sum_{k\in\mathbb{Z}}\overline{\varphi}(k)e^{i(n-k)\lambda} \text{ in } L^2(\mathbb{T},\sigma).$$

Therefore, we may conclude the X_n, $n \in \mathbb{Z}$ has a moving average representation of the form

$$X_n = \sum_{k\in\mathbb{Z}}\overline{\hat{\varphi}(k)}\xi_{n-k},$$

for some white noise sequence $(\xi_n)_{n\in\mathbb{Z}}$ contained in $L(X)$. We know that our white noise sequence is contained in $L(X)$ because $H^2(\mathbb{T})$ functions do not vanish on a set of positive Lebesgue measure. Furthermore, we know that W is actually an isomorphism and $\xi_n = W^{-1}(e^{in\lambda})$, for all $n \in \mathbb{Z}$.

Theorem 1.8.4. *Suppose that X_n, $n \in \mathbb{Z}$ is a weakly stationary sequence with spectral measure $dF_X(e^{i\lambda}) = |\varphi(e^{i\lambda})|^2 \, d\sigma(e^{i\lambda})$, where $\varphi \in H^2(\mathbb{T})$. Then, X_n, $n \in \mathbb{Z}$ may be represented as a one-sided moving average sequence; that is,*

$$X_n = \sum_{k\in\mathbb{N}}\overline{\hat{\varphi}(k)}\,\xi_{n-k},$$

in $L^2(\Omega,\mathcal{F},P)$, where ξ_n, $n \in \mathbb{Z}$ is a white noise sequence contained in $L(X)$.

We now examine proper one-sided moving representations. Suppose that X_n, $n \in \mathbb{Z}$ has a one-sided moving average representation. Therefore,

$$X_n = \sum_{k\in\mathbb{N}}a_k\,\xi_{n-k},$$

in $L^2(\Omega,\mathcal{F},P)$, for all $n \in \mathbb{Z}$, where $(a_n)_{n\in\mathbb{N}}$ is a complex sequence with $\sum_{n\in\mathbb{N}}|a_n|^2 < \infty$ and ξ_n, $n \in \mathbb{Z}$ is a white noise sequence in $L^2(\Omega,\mathcal{F},P)$.

Now, define

$$L(\xi : n) = \overline{\text{span}}\,\{\xi_k : k \in \mathbb{Z}, k \leq n\}$$

and

$$L(X : n) = \overline{\text{span}}\,\{X_k : k \in \mathbb{Z}, k \leq n\}.$$

It is of interest to know when a one-sided moving average is a **proper one-sided moving average**. A one-sided moving average is called **proper** when $L(\xi : n) = L(X : n)$, for all $n \in \mathbb{Z}$.

Define the linear transformation $W^\# : L(\xi) \to L^2(\mathbb{T},\sigma)$ by $W^\#(\xi_n) = e^{-in\lambda}$. $W^\#$ is clearly an isomorphism. From the moving average representation, we see that

$$W^\#(X_n) = e^{-in\lambda}f(e^{i\lambda}),$$

where $f(e^{i\lambda}) = \sum_{k\in\mathbb{N}}a_k e^{ik\lambda}$ in $L^2(\mathbb{T},\sigma)$. From this representation, we see that $f \in H^2(\mathbb{T})$ and therefore has an inner-outer representation. That is, $f = IO$, where I is an inner function and O is an outer function. Using the fact that the following subspaces are simply invariant under multiplication by the coordinate function, we get from Theorem 3.2.2 that

$$W^{\#}(L(X:n)) = \overline{\operatorname{span}}\left\{e^{-ik\lambda}f(e^{i\lambda}) : k \in \mathbb{Z}, k \le n\right\} = e^{-in\lambda}I(e^{i\lambda})H^2(\mathbb{T}).$$

Similarly, we get that

$$W^{\#}(L(\xi:n)) = \overline{\operatorname{span}}\left\{e^{-ik\lambda} : k \in \mathbb{Z}, k \le n\right\} = e^{-in\lambda}H^2(\mathbb{T}).$$

These observations show that X_n, $n \in \mathbb{Z}$ has a proper one-sided moving average representation precisely when f is an outer function.

We will now record this observation as a theorem.

Theorem 1.8.5. *Suppose that X_n, $n \in \mathbb{Z}$ has a one-sided moving average representation. Therefore,*

$$X_n = \sum_{k \in \mathbb{N}} a_k\, \xi_{n-k},$$

in $L^2(\Omega, \mathcal{F}, P)$, for all $n \in \mathbb{Z}$, where $(a_n)_{n\in\mathbb{N}}$ is a complex sequence with $\sum_{n\in\mathbb{N}} |a_n|^2 < \infty$ and ξ_n, $n \in \mathbb{Z}$ is a white noise sequence in $L^2(\Omega, \mathcal{F}, P)$. Then, this is a proper one-sided moving average representation if and only if the sequence $(a_n)_{n\in\mathbb{N}}$ are the coefficients of an outer function in $H^2(\mathbb{T})$.

The following theorem follows from Theorem 1.8.4 and Theorem 1.8.5.

Theorem 1.8.6. *Suppose that X_n, $n \in \mathbb{Z}$ is a weakly stationary sequence with spectral measure $dF_X(e^{i\lambda}) = |\varphi(e^{i\lambda})|^2\, d\sigma(e^{i\lambda})$, where $\varphi \in H^2(\mathbb{T})$. Then, if φ is an outer function, then X_n, $n \in \mathbb{Z}$ may be represented as a proper one-sided moving average sequence.*

Since inner functions have modulus one $[\sigma]$-a.e., the inner factor has no effect on the spectral measure. For this reason, if we only use the outer factor to construct the one-sided moving average, our representation will be proper. Combining these observations with Theorem 1.8.6, we get the following theorem.

Theorem 1.8.7. *Suppose that X_n, $n \in \mathbb{Z}$ is a weakly stationary sequence with spectral measure $dF_X(e^{i\lambda}) = |\varphi(e^{i\lambda})|^2\, d\sigma(e^{i\lambda})$, where $\varphi \in H^2(\mathbb{T})$. Then, X_n, $n \in \mathbb{Z}$ may be represented as a proper one-sided moving average sequence.*

Note: Outer functions are unique up to a constant multiple of modulus one.

1.9 Regular and Singular Sequences

Let X_n, $n \in \mathbb{Z}$ be a weakly stationary sequence. Let $L(X)$ and $L(X : n)$ be as defined previously. Define

$$L(X : -\infty) = \bigcap_n L(X : n). \qquad (1.4)$$

Note that $L(X : -\infty) \subseteq L(X : n) \subseteq L(X : n+1) \subset L(X)$. We will call X_n, $n \in \mathbb{Z}$, deterministic or singular if $L(X : -\infty) = L(X)$ and purely non-deterministic or regular if $L(X : -\infty) = \{0\}$. Before we revisit our examples of weakly stationary sequences and determine whether they are regular or singular, let us make some observations. We will state these observations as propositions and leave them to the reader to prove.

Proposition 1.9.1. *If $L(X : n) = L(X : n-1)$ for all $n \in \mathbb{Z}$, then X_n, $n \in \mathbb{Z}$ is singular.*

Proposition 1.9.2. *If $X_n \in L(X : n - 1)$ for all $n \in \mathbb{Z}$, then $L(X : n) = L(X : n - 1)$ for all $n \in \mathbb{Z}$.*

Proposition 1.9.3. *If there exists a $k \in \mathbb{Z}$ such that $L(X : k) = L(X : k-1)$, then $L(X : n) = L(X : n - 1)$ for all $n \in \mathbb{Z}$.*

1.10 Examples of Regular and Singular Sequences

We now revisit our old examples and analyze them in the context of regularity and singularity. Recall that L_0^2 denotes the set of all $\zeta \in L^2(\Omega, \mathcal{F}, P)$ with the property that $E\zeta = 0$.

Example 1.10.1. *Let $\zeta \in L_0^2$, $\gamma \in [-\pi, \pi)$ and define $X_n = \zeta e^{in\gamma}$, $n \in \mathbb{Z}$. We recall the following about this example.*

- X_n, $n \in \mathbb{Z}$ is a weakly stationary sequence.

- X_n, $n \in \mathbb{Z}$ has covariance function $r(n) = e^{in\gamma} E|\zeta|^2$, $n \in \mathbb{Z}$.

- X_n, $n \in \mathbb{Z}$ has spectral measure F_X that is a discrete measure concentrated on the point $e^{i\gamma}$ with mass $F_X(\{e^{i\gamma}\}) = E|\zeta|^2$.

Now, since $X_n = e^{i\gamma} X_{n-1} \in L(X : n - 1)$, it follows from Proposition 1.9.1 and Proposition 1.9.2 that X_n, $n \in \mathbb{Z}$ is singular.

Example 1.10.2. *Let $Z_1, ..., Z_N$ be an orthogonal collection of random variables in L_0^2 and let $\gamma_1, ..., \gamma_N \in [-\pi, \pi)$. Now, define $X_n = \sum_{k=1}^{N} Z_k e^{in\gamma_k}$, $n \in \mathbb{Z}$.*

We recall the following about this example.

- X_n, $n \in \mathbb{Z}$ *is a weakly stationary sequence.*

- X_n, $n \in \mathbb{Z}$ *has covariance function $r(n) = \sum_{k=1}^{N} e^{in\gamma_k} E|Z_k|^2$, $n \in \mathbb{Z}$.*

- X_n, $n \in \mathbb{Z}$ *has spectral measure F_X that is a discrete measure concentrated at the points $e^{i\gamma_1}, ..., e^{i\gamma_N}$ with masses $F_X(\{e^{i\gamma_k}\}) = E|Z_k|^2$.*

We leave it to the reader to show that there exists constants $\alpha_1, ..., \alpha_N$ such that $X_n = \sum_{k=1}^{N} \alpha_k X_{n-k}$. Therefore, $X_n \in L(X : n-1)$. Therefore, by Proposition 1.9.1 and Proposition 1.9.2, X_n, $n \in \mathbb{Z}$ is singular.

Example 1.10.3 (White Noise). *Take X_n, $n \in \mathbb{Z}$ to be an orthonormal sequence of random variables in L_0^2. We recall the following about this example.*

- X_n, $n \in \mathbb{Z}$ *is a weakly stationary sequence.*

- X_n, $n \in \mathbb{Z}$ *has covariance function $r(n) = \delta_{n0}$, $n \in \mathbb{Z}$.*

- X_n, $n \in \mathbb{Z}$ *has a spectral measure that is absolutely continuous with respect to normalized Lebesgue measure with density function $f_X(e^{i\lambda}) = 1$. That is, $dF_X(e^{i\lambda}) = d\sigma(e^{i\lambda})$.*

We will show that X_n, $n \in \mathbb{Z}$ is regular. To see this, first note that X_n, $n \in \mathbb{Z}$ is an orthonormal basis for $L(X)$. Therefore, X_k, $k \in \mathbb{Z}$, with $k \leq n$ forms an orthonormal basis for $L(X : n)$. Therefore, if $f \in L(X : -\infty)$, then $f = \sum_{k=-\infty}^{n} c_{n,k} X_k$, for all $n \in \mathbb{Z}$. Similarly, since $f \in L(X)$, $f = \sum_{k=-\infty}^{\infty} c_k X_k$. Now, since this representation is unique, it follows that $c_k = 0$ for all $k \in \mathbb{Z}$. Therefore, $f = 0$ and $L(X : -\infty) = \{0\}$, as desired.

Example 1.10.4 (One-sided Moving Average Sequence). *Let $a_k \in \mathbb{C}$, $k \in \mathbb{N}$ with $\sum_{k=0}^{\infty} |a_k|^2 < \infty$ and let ξ_n, $n \in \mathbb{Z}$ be a white noise sequence. Now, take $X_n = \sum_{k=0}^{\infty} a_k \xi_{n-k}$, $n \in \mathbb{Z}$. We recall the following about this example.*

- X_n, $n \in \mathbb{Z}$ *is a weakly stationary sequence.*

- X_n, $n \in \mathbb{Z}$ *has covariance function $r(n) = \sum_{k=0}^{\infty} a_k \bar{a}_{k-n} = \sum_{k=0}^{\infty} a_{n+k} \bar{a}_k$, $n \in \mathbb{Z}$.*
  In this expression, we take each coefficient with a negative index to be zero.

- X_n, $n \in \mathbb{Z}$ has spectral measure that is absolutely continuous with respect to normalized Lebesgue measure with density function $f_X(e^{i\lambda}) = \left| \sum_{k=0}^{\infty} \overline{a}_k e^{ik\lambda} \right|^2$.

That is, $dF_X(e^{i\lambda}) = \left| \sum_{k=0}^{\infty} \overline{a}_k e^{ik\lambda} \right|^2 d\sigma(e^{i\lambda})$.

By Theorem 1.8.7, X_n, $n \in \mathbb{Z}$ may be represented as a proper one-sided moving average, with associated white noise sequence, say $(\epsilon_n)_{n \in \mathbb{Z}}$. Being proper gives us that $L(X : n) = L(\epsilon : n)$ for all $n \in \mathbb{Z}$. Therefore, $L(X : -\infty) = L(\epsilon : -\infty)$ and from our last example, $L(\epsilon : -\infty) = \{0\}$. Therefore, X_n, $n \in \mathbb{Z}$ is regular.

Example 1.10.5 (Moving Average Sequence). *Let* $a_k \in \mathbb{C}$, $k \in \mathbb{Z}$ *with* $\sum_{k=-\infty}^{\infty} |a_k|^2 < \infty$ *and let* ξ_n, $n \in \mathbb{Z}$ *be a white noise sequence. Now, take* $X_n = \sum_{k=-\infty}^{\infty} a_k \xi_{n-k}$, $n \in \mathbb{Z}$. *We recall the following about this example.*

- X_n, $n \in \mathbb{Z}$ is a weakly stationary sequence.

- X_n, $n \in \mathbb{Z}$ has covariance function $r(n) = \sum_{k=-\infty}^{\infty} a_k \overline{a}_{k-n} = \sum_{k=-\infty}^{\infty} a_{n+k} \overline{a}_k$, $n \in \mathbb{Z}$.

- X_n, $n \in \mathbb{Z}$ has spectral measure that is absolutely continuous with respect to normalized Lebesgue measure with density function $f_X(e^{i\lambda}) = \left| \sum_{k=-\infty}^{\infty} \overline{a}_k e^{ik\lambda} \right|^2$. That is, $dF_X(e^{i\lambda}) = \left| \sum_{k=-\infty}^{\infty} \overline{a}_k e^{ik\lambda} \right|^2 d\sigma(e^{i\lambda})$.

Define the linear transformation $W : L(\xi) \to L^2(\mathbb{T}, \sigma)$ by $W(\xi_n) = e^{-in\lambda}$. W is clearly an isomorphism. From the moving average representation, we see that
$$W(X_n) = e^{-in\lambda} g(e^{i\lambda}),$$
where $g(e^{i\lambda}) = \sum_{k \in \mathbb{Z}} a_k e^{ik\lambda}$ in $L^2(\mathbb{T})$. Therefore,
$$W(L(X : n)) = \overline{\text{span}} \left\{ e^{-ik\lambda} g(e^{i\lambda}) : k \in \mathbb{Z}, k \leq n \right\}$$
for all $n \in \mathbb{Z}$. These subspaces of $L^2(\mathbb{T}, \sigma)$ are invariant under multiplication by the coordinate function. As such, they have a specific form depending on g. See Section 3.2 for the details.

If g does not have the modulus of an $H^2(\mathbb{T})$ function, then $W(L(X : n))$ is a doubly invariant subspace of $L^2(\mathbb{T}, \sigma)$ for each $n \in \mathbb{Z}$ and $W(L(X : n)) = \mathbf{1}_E L^2(\mathbb{T}, \sigma)$, for each $n \in \mathbb{Z}$, where $E = \{e^{i\lambda} : g(e^{i\lambda}) \neq 0\}$. Therefore, $L(X : n) = L(X : n - 1)$ for all $n \in \mathbb{Z}$. Therefore, by Proposition 1.9.1, X_n, $n \in \mathbb{Z}$ is singular.

If g has the modulus of an $H^2(\mathbb{T})$ function, then so does $\sqrt{f_X}$. Therefore, we may write $f_X(e^{i\lambda}) = \left|\varphi(e^{i\lambda})\right|^2$, where $\varphi \in H^2(\mathbb{T})$. Therefore, by Theorem 1.8.4, X_n, $n \in \mathbb{Z}$ may be represented as a one-sided moving average and therefore, by our last example, X_n, $n \in \mathbb{Z}$ is regular.

1.11   The Wold Decomposition

Let X_n, $n \in \mathbb{Z}$ be a weakly stationary sequence and let

$$W_n = L(X : n) \ominus L(X : n - 1)$$

for all $n \in \mathbb{Z}$.

Lemma 1.11.1. $L(X : n) = \sum_{k=0}^{\infty} \oplus W_{n-k} \bigoplus L(X : -\infty).$

Proof. Henceforth, P_M will denote the projection on $L(X)$ onto the (closed) subspace M.

$$
\begin{aligned}
P_{L(X:n)} &= (P_{L(X:n)} - P_{L(X:n-1)}) + P_{L(X:n-1)} \\
&= P_{W_n} + P_{L(X:n-1)} \\
&= P_{W_n} + (P_{L(X:n-1)} - P_{L(X:n-2)}) + P_{L(X:n-2)} \\
&= P_{W_n} + P_{W_{n-1}} + P_{L(X:n-2)} \\
&= \cdots \\
&= \sum_{k=0}^{l} P_{W_{n-k}} + P_{L(X:n-l-1)}
\end{aligned}
$$

Letting l go to infinity, we get

$$P_{L(X:n)} = \sum_{k=0}^{\infty} P_{W_{n-k}} + P_{L(X:-\infty)}. \qquad (1.5)$$

It then follows that

$$L(X : n) = \sum_{k=0}^{\infty} \oplus W_{n-k} \bigoplus L(X : -\infty) \qquad (1.6)$$

as desired. $\square$

Lemma 1.11.2. $L(X) = \sum_{k=-\infty}^{\infty} \oplus W_k \bigoplus L(X : -\infty).$

Proof. We may rewrite (1.5) as

$$P_{L(X:n)} = \sum_{k=-\infty}^{n} P_{W_k} + P_{L(X:-\infty)}. \tag{1.7}$$

Now, letting n go to infinity, we get

$$P_{L(X)} = \sum_{k=-\infty}^{\infty} P_{W_k} + P_{L(X:-\infty)}. \tag{1.8}$$

It then follows that

$$L(X) = \sum_{k=-\infty}^{\infty} \oplus W_k \bigoplus L(X:-\infty) \tag{1.9}$$

as desired. $\qquad\square$

Lemma 1.11.3. *The following are equivalent:*

1. $L(X:-\infty) = \{0\}$.

2. $\lim_{l \to -\infty} P_{L(X:l)} X_n = 0$ *for all* $n \in \mathbb{Z}$.

3. $X_n = \sum_{k=0}^{\infty} a_{n,k} \nu_{n-k}$ *for all* $n \in \mathbb{Z}$, *where* ν_n, $n \in \mathbb{Z}$ *is an orthogonal sequence in* $L(X)$.

Proof. $(1. \Rightarrow 2.)$ $P_{L(X:l)} \downarrow P_{L(X:-\infty)} = P_{\{0\}}$.

$(2. \Rightarrow 3.)$ By Lemma 1.11.1, $L(X:n) = \sum_{k=0}^{\infty} \oplus W_{n-k} \bigoplus L(X:-\infty)$. There-

fore, $X_n = \sum_{k=0}^{\infty} a_{n,k} \nu_{n-k}$, where $\nu_k \in W_k$ for all k and hence forms an orthog-

onal sequence in $L(X)$.

$(3. \Rightarrow 1.)$ Let

$$N_n = \overline{\text{span}} \left\{ \frac{\nu_k}{\|\nu_k\|} : k \le n \right\}. \tag{1.10}$$

Therefore, $X_m \in N_n$ for all $m \le n$ and hence $L(X:-\infty) \subseteq N_{-\infty} := \cap_n N_n$.
Let $u \in N_{-\infty}$. Then

$$u = \sum_{k=-\infty}^{n} c_k \eta_k \qquad \forall n \in \mathbb{Z},$$

where $\eta_k = \frac{\nu_k}{\|\nu_k\|}$. Note that c_k does not depend on n. To see this, note that the
η_k form an orthonormal sequence and as such each representation is unique.
Therefore, given any value n, the representation up to $n+1$ can be found by
making $c_{n+1} = 0$ and, therefore, it follows that $c_k = 0$ for all $k \in \mathbb{Z}$. Therefore,
$u = 0$ and hence both $N_{-\infty} = \{0\}$ and $L(X:-\infty) = \{0\}$. $\qquad\square$

We now give the main result for this section.

Theorem 1.11.1 (Wold Decomposition). *Suppose X_n, $n \in \mathbb{Z}$ is a weakly stationary sequence. Then, X_n may be written as*

$$X_n = X_n^r + X_n^s, \qquad n \in \mathbb{Z},$$

where

1. *X_n^r, $n \in \mathbb{Z}$ is a weakly stationary regular sequence.*

2. *X_n^s, $n \in \mathbb{Z}$ is a weakly stationary singular sequence.*

3. *$L(X^s : n) \subseteq L(X : n)$, $L(X^r : n) \subseteq L(X : n)$ and $L(X^r : n) \perp L(X^s : n)$ for all $n \in \mathbb{Z}$.*

4. *$L(X : n) = L(X^r : n) \oplus L(X^s : n)$ for all $n \in \mathbb{Z}$.*

5. *The decomposition satisfying these conditions is unique.*

Before proving the Wold Decomposition, we state and prove two lemmas.

Lemma 1.11.4. $U P_{L(X:n)} = P_{L(X:n+1)} U$ on $L(X)$.

Proof. Let $Y \in L(X)$. Then, Y has a unique decomposition

$$Y = P_{L(X:n)}Y + \left(Y - P_{L(X:n)}Y\right).$$

Note that $P_{L(X:n)}Y \in L(X : n)$ and $\left(Y - P_{L(X:n)}Y\right) \in L(X) \ominus L(X : n)$. Now, applying U to both sides, we get that

$$UY = U P_{L(X:n)}Y + \left(UY - U P_{L(X:n)}Y\right).$$

This is a unique decomposition of UY with $U P_{L(X:n)}Y \in L(X : n + 1)$ and $\left(UY - U P_{L(X:n)}Y\right) \in L(X) \ominus L(X : n+1)$. Note that using the same approach as that with Y, we get the unique decomposition

$$UY = P_{L(X:n+1)}UY + \left(UY - P_{L(X:n+1)}UY\right),$$

where $P_{L(X:n+1)}UY \in L(X : n+1)$ and $\left(UY - P_{L(X:n+1)}UY\right) \in L(X) \ominus L(X : n + 1)$. The uniqueness of these decompositions shows that

$$U P_{L(X:n)}Y = P_{L(X:n+1)}UY.$$

Therefore, $U P_{L(X:n)} = P_{L(X:n+1)}U$ on $L(X)$, as desired. $\square$

Lemma 1.11.5. $U P_{L(X:-\infty)} = P_{L(X:-\infty)}U$ on $L(X)$.

Proof. Recall that $P_{L(X:n)}$ converges to $P_{L(X:-\infty)}$ strongly. Therefore, $P_{L(X:n)}U$ converges to $P_{L(X:-\infty)}U$ strongly. Since U is unitary, it also follows that $U P_{L(X:n)}$ converges to $U P_{L(X:-\infty)}$ strongly. Hence, for $Y \in L(X)$, we have

$$\|U P_{L(X:-\infty)}Y - P_{L(X:-\infty)}UY\|_{L(X)}$$

$$= \|UP_{L(X:-\infty)}Y - UP_{L(X:n)}Y + P_{L(X:n+1)}UY - P_{L(X:-\infty)}UY\|_{L(X)}$$

$$\leq \|UP_{L(X:-\infty)}Y - UP_{L(X:n)}Y\|_{L(X)} + \|P_{L(X:n+1)}UY - P_{L(X:-\infty)}UY\|_{L(X)}$$

$$\longrightarrow 0 \text{ as } n \longrightarrow \infty.$$

Therefore, $UP_{L(X:-\infty)} = P_{L(X:-\infty)}U$ on $L(X)$. $\qquad\qquad\Box$

We are now ready to give the proof of the Wold Decomposition.

Proof of Theorem 1.11.1. Let $X_n^s = P_{L(X:-\infty)}X_n$ and $X_n^r = X_n - X_n^s$ for all $n \in \mathbb{Z}$. It then follows from their definitions that:

1. $X_n = X_n^r + X_n^s$ for all $n \in \mathbb{Z}$.

2. $X_m^r \perp X_n^s$ for all $m, n \in \mathbb{Z}$. Therefore, $L(X^r : n) \perp L(X^s : n)$ for all $n \in \mathbb{Z}$.

3. $L(X : n) \subseteq L(X^r : n) + L(X^s : n)$ for all $n \in \mathbb{Z}$.

Recalling Lemma 2, we also see that $L(X^s : n) \subseteq L(X : n)$ and $L(X^r : n) \subseteq L(X : n)$. It then follows that $L(X^r : n) + L(X^s : n) \subseteq L(X : n)$ for all $n \in \mathbb{Z}$. Putting this observation together with item number 3 above, we get that

$$L(X : n) = L(X^r : n) \oplus L(X^s : n) \text{ for all } n \in \mathbb{Z}.$$

The following two observations can be used to show that X_n^r and X_n^s, $n \in \mathbb{Z}$ are stationary sequences. Let U be the unitary operator on $L(X)$ that takes X_n to X_{n+1}. We observe that

1. $UX_n^s = UP_{L(X:-\infty)}X_n = P_{L(X:-\infty)}UX_n = P_{L(X:-\infty)}X_{n+1} = X_{n+1}^s$.

2. $UX_n^r = U(X_n - X_n^s) = UX_n - UX_n^s = X_{n+1} - X_{n+1}^s = X_{n+1}^r$.

Note that in observation 1, we used the commuting condition observed prior to this proof and in observation 2, we used observation 1.

We will now show that X_n^r, $n \in \mathbb{Z}$ is regular, and X_n^s, $n \in \mathbb{Z}$ is singular. To see that X_n^s, $n \in \mathbb{Z}$ is singular, note that $L(X^s : n) = L(X : n) \cap L(X : -\infty) = L(X : -\infty)$. This follows from the definition of X_n^s and Lemma 2. It then follows that $L(X^s : -\infty) = L(X^s)$. That is, X_n^s, $n \in \mathbb{Z}$ is singular. To see that X_n^r, $n \in \mathbb{Z}$ is regular, note by our previous observation that

$$L(X^r : n) = L(X : n) \ominus L(X^s : n) \text{ for all } n \in \mathbb{Z}.$$

Our previous observation then gives

$$L(X^r : n) = L(X : n) \ominus L(X : -\infty) \text{ for all } n \in \mathbb{Z}.$$

It then follows that $L(X^r : -\infty) = \{0\}$. That is, X_n^r, $n \in \mathbb{Z}$ is regular, as desired.

It is left to show that this decomposition is unique. To see this, we employ property 3 of the Wold Decomposition, which states that if X_n has another Wold Decomposition $X_n = \tilde{X}_n^r + \tilde{X}_n^s$ then $L(\tilde{X}^s : n) \subseteq L(X : n)$ and therefore, $L(\tilde{X}^s : -\infty) \subseteq L(X : -\infty)$. Since $\tilde{X}_n^s$ is singular, we may conclude that $L(\tilde{X}^s) \subseteq L(X : -\infty)$. Now using property 4 of the Wold Decomposition along with the fact that $\tilde{X}_n^r$ is regular, we see that $L(\tilde{X}^s) = L(X : -\infty)$. Therefore, the following calculation yields the uniqueness of this decomposition.

$$X_n^s = P_{L(X:-\infty)}X_n = P_{L(\tilde{X}^s)}X_n = \tilde{X}_n^s.$$

$\square$

1.12   The Theory of Regular Sequences

In this section, we present some theorems that give a complete characterization of regular sequences, both in terms of their moving average representation and their spectral measure.

Theorem 1.12.1. *If a weakly stationary sequence X_n, $n \in \mathbb{Z}$ is regular, then X_n, $n \in \mathbb{Z}$ has a proper one-sided moving average representation.*

Proof. Let $\xi_n = \dfrac{X_n - P_{L(X:n-1)}X_n}{\|X_n - P_{L(X:n-1)}X_n\|}$ and let U be the unitary operator defined on $L(X)$ that takes X_n to X_{n+1} then

$$
\begin{aligned}
\xi_{n+1} &= \frac{X_{n+1} - P_{L(X:n)}X_{n+1}}{\|X_{n+1} - P_{L(X:n)}X_{n+1}\|} \\
&= \frac{UX_n - P_{L(X:n)}UX_n}{\|UX_n - P_{L(X:n)}UX_n\|} \\
&= \frac{UX_n - UP_{L(X:n-1)}X_n}{\|UX_n - UP_{L(X:n-1)}X_n\|} \\
&= \frac{U(X_n - P_{L(X:n-1)}X_n)}{\|U(X_n - P_{L(X:n-1)}X_n)\|} \\
&= U\xi_n.
\end{aligned}
$$

The last equality follows from the fact that U is an isometry. By construction, $\xi_n \in L(X : n)$ and ξ_n is orthogonal to $L(X : n - 1)$. It follows that $L(\xi : n) \subseteq L(X : n)$. The fact that X_n is a regular sequence implies that ξ_n, $n \in \mathbb{Z}$ is an orthonormal basis for $L(X)$. Therefore, $L(\xi : n) = L(X : n)$.

The fact that ξ_n, $n \in \mathbb{Z}$ is an orthonormal basis for $L(X)$ implies that there exists a sequence of complex numbers $(a_k)_{k=0}^{\infty}$ with $\displaystyle\sum_{k=0}^{\infty} |a_k|^2 < \infty$ such that

$$X_0 = \sum_{k=0}^{\infty} a_k \xi_{-k}. \tag{1.11}$$

This implies that

$$X_n = U^n X_0 = \sum_{k=0}^{\infty} a_k U^n \xi_{-k} = \sum_{k=0}^{\infty} a_k \xi_{n-k},$$

as desired.   □

Notice that this is an improvement from Lemma 1.11.3, since here the a_k do not depend on n as they did in Lemma 1.11.3.

The following proposition was already presented as an example, but is included here to both complement the previous theorem and for ease of reference.

Proposition 1.12.1. *If X_n, $n \in \mathbb{Z}$ has a one-sided moving average representation, then X_n, $n \in \mathbb{Z}$ is regular.*

Combining some earlier results, we get the following results.

Theorem 1.12.2. *If a weakly stationary sequence X_n, $n \in \mathbb{Z}$ is regular, then the spectral measure for this sequence is absolutely continuous with respect to normalized Lebesgue measure with density $f_X(e^{i\lambda}) = |\varphi(e^{i\lambda})|^2$, where $\varphi \in H^2(\mathbb{T})$. Furthermore, one may choose φ to be an outer function.*

Theorem 1.12.3. *If a weakly stationary sequence X_n, $n \in \mathbb{Z}$ has a spectral measure that is absolutely continuous with respect to normalized Lebesgue measure with density $f_X(e^{i\lambda}) = |\varphi(e^{i\lambda})|^2$, where $\varphi \in H^2(\mathbb{T})$, then X_n, $n \in \mathbb{Z}$ is regular.*

1.13   The Concordance Theorem

We now present a theorem that gives a decomposition of the spectral measure of a weakly stationary sequence with respect to its Wold Decomposition.

Theorem 1.13.1 (Concordance Theorem). *Let X_n, $n \in \mathbb{Z}$ be a weakly stationary sequence, that is not singular, with Wold Decomposition $X_n = X_n^r + X_n^s$. We then have the following:*

1. *$F_X = F_{X^r} + F_{X^s}$, where F_X, F_{X^r} and F_{X^s} are the spectral measures of X, X^r and X^s, respectively.*

    2. F_{X^r} is absolutely continuous with respect to normalized Lebesgue measure with density $f_{X^r}(e^{i\lambda}) = |\varphi(e^{i\lambda})|^2$, where $\varphi \in H^2(\mathbb{T})$.

    3. F_{X^s} and Lebesgue measure are mutually singular.

Note: If X_n, $n \in \mathbb{Z}$ is singular, then its spectral measure may be absolutely continuous with respect to normalized Lebesgue measure, as illustrated in Example 1.10.5.

Proof. To prove 1, note that

$$
\begin{aligned}
r_X(n) &= EX_n\overline{X_0} \\
&= E(X_n^r + X_n^s)\overline{(X_0^r + X_0^s)} \\
&= EX_n^r\overline{X_0^r} + EX_n^s\overline{X_0^s} \\
&= r_{X^r}(n) + r_{X^s}(n),
\end{aligned}
$$

where r_X, r_{X^r} and r_{X^s} are the covariance functions of X, X^r and X^s, respectively. It then follows from the uniqueness of the spectral measure that $F_X = F_{X^r} + F_{X^s}$, where F_X, F_{X^r} and F_{X^s} are the spectral measures of X, X^r and X^s, respectively.

    Part 2 follows directly from Theorem 1.12.2, above. To see part 3, we make a few observations. First, we note from the Wold Decomposition that $X_0 = X_0^r + X_0^s$. Now, using the spectral representation, we get that

$$X_0 = \int_{[-\pi,\pi)} d\mathcal{Z}_X(e^{i\lambda}) \tag{1.12}$$

and that there exists g_r and g_s in $L^2(\mathbb{T}, F_X)$ such that

$$X_0^r = \int_{[-\pi,\pi)} g_r(e^{i\lambda}) \, d\mathcal{Z}_X(e^{i\lambda}) \tag{1.13}$$

and

$$X_0^s = \int_{[-\pi,\pi)} g_s(e^{i\lambda}) \, d\mathcal{Z}_X(e^{i\lambda}). \tag{1.14}$$

If we use (1.12), (1.13) and (1.14), we get

$$
\begin{aligned}
\int_{[-\pi,\pi)} dF_X(e^{i\lambda}) = \|X_0\|^2 &= \|X_0^r\|^2 + \|X_0^s\|^2 \\
&= \int_{[-\pi,\pi)} (|g_r(e^{i\lambda})|^2 + |g_s(e^{i\lambda})|^2) \, dF_X(e^{i\lambda}).
\end{aligned}
$$

By the uniqueness of the spectral measure, F_X, we see that

$$|g_r(e^{i\lambda})|^2 + |g_s(e^{i\lambda})|^2 = 1 \ [F_X]\text{-a.e.} \tag{1.15}$$

Now, by an earlier observation, we have that

$$X_n^r = \int_{[-\pi,\pi)} e^{in\lambda} g_r(e^{i\lambda}) \, d\mathcal{Z}_X(e^{i\lambda})$$

and

$$X_m^s = \int_{[-\pi,\pi)} e^{im\lambda} g_s(e^{i\lambda}) \, d\mathcal{Z}_X(e^{i\lambda}).$$

Recalling that $X_n^r \perp X_m^s$ for all $m, n \in \mathbb{Z}$, it follows that

$$\int_{[-\pi,\pi)} e^{ik\lambda} g_r(e^{i\lambda}) \overline{g_s(e^{i\lambda})} \, dF_X(e^{i\lambda}) = 0,$$

for all $k \in \mathbb{Z}$. Therefore,

$$g_r(e^{i\lambda}) \overline{g_s(e^{i\lambda})} = 0 \ [F_X]\text{-a.e.} \tag{1.16}$$

Now putting (1.15) and (1.16) together, we get

$$0 = |g_r(e^{i\lambda}) \overline{g_s(e^{i\lambda})}|^2 = |g_r(e^{i\lambda})|^2 |g_s(e^{i\lambda})|^2 = |g_r(e^{i\lambda})|^2 (1 - |g_r(e^{i\lambda})|^2) \ [F_X]\text{-a.e.}$$

It follows from this that $|g_r(e^{i\lambda})|^2$ and $|g_s(e^{i\lambda})|^2$ take values 0 or 1 $[F_X]$-a.e. Since X_n^r and X_n^s are both weakly stationary, we may apply the spectral representation to get the following equations

$$dF_{X^s}(e^{i\lambda}) = |g_s(e^{i\lambda})|^2 \, dF_X(e^{i\lambda}) \tag{1.17}$$

and

$$dF_{X^r}(e^{i\lambda}) = |g_r(e^{i\lambda})|^2 \, dF_X(e^{i\lambda}). \tag{1.18}$$

If we now let $R = \{e^{i\lambda} : g_r(e^{i\lambda}) = 1\}$ and $S = \{e^{i\lambda} : g_s(e^{i\lambda}) = 1\}$, it follows that R and S are disjoint with F_{X^r} concentrated on R and F_{X^s} concentrated on S. That is, F_{X^r} and F_{X^s} are mutually singular. It follows from part 2 of this theorem, which we already proved, that $f_{X^r}(e^{i\lambda}) = |\phi(e^{i\lambda})|^2$, where $\phi \in H^2(\mathbb{T})$. Therefore,

$$|\phi(e^{i\lambda})|^2 \, d\sigma(e^{i\lambda}) = |g_r(e^{i\lambda})|^2 \, dF_X(e^{i\lambda}),$$

which shows that R has full Lebesgue measure since $\phi \neq 0 \ [\sigma]$-a.e. Therefore, S has Lebesgue measure zero. Hence, we may conclude that F_{X^s} and Lebesgue measure are mutually singular. $\qquad\square$

Before moving on, we point out that we can get the spectral form of the Wold Decomposition

$$X_n^{(1)} = \int_{[-\pi,\pi)} e^{in\lambda} \, d\mathcal{Z}_{X^r}(e^{i\lambda}) \text{ and } X_n^{(2)} = \int_{[-\pi,\pi)} e^{in\lambda} \, d\mathcal{Z}_{X^s}(e^{i\lambda}),$$

where $\mathcal{Z}_{X^r}(\Delta) = \mathcal{Z}_X(\Delta \cap R)$ and $\mathcal{Z}_{X^s}(\Delta) = \mathcal{Z}_X(\Delta \cap S)$. This gives the concordance between the Wold and Cramer decompositions.

1.14   Maximal Factors and Innovation Processes

In this section, we introduce some new terminology based on an important observation. As we know, if X_n, $n \in \mathbb{Z}$ is regular, then the spectral measure for this sequence is absolutely continuous with respect to normalized Lebesgue measure with density $f(e^{i\lambda}) = |\varphi(e^{i\lambda})|^2$, where $\varphi \in H^2(\mathbb{T})$. Furthermore, it may be represented as a one-sided moving average in the following way:

$$X_n = \sum_{k=0}^{\infty} \overline{\hat{\varphi}}(k)\xi_{n-k},$$

for some white noise sequence ξ_n, $n \in \mathbb{Z}$ in $L(X)$. We observe that

$$\left\| X_1 - P_{L(\xi,0)}X_1 \right\|_{L(X)} = |\hat{\varphi}(0)| .$$

This value is maximal if φ is an outer function. For this reason, we will often refer to the outer function associated with a given regular sequence as the maximal factor. The white noise sequence associated with the maximal factor will be called the innovation process. Recall that the maximal factor for a given regular sequence is unique up to a constant multiple of modulus one.

Theorem 1.14.1. *Let X_n, $n \in \mathbb{Z}$ be a weakly stationary regular sequence with maximal factor ϕ and innovation process ν_n, $n \in \mathbb{Z}$. Finally, let $\mathcal{J}$: $L^2(\mathbb{T}, F_X) \to L(X)$ be our canonical isomorphism that takes $e^{in\lambda}$ to X_n. We then have the following:*

1. $\mathcal{J}^{-1}(\nu_n) = e^{in\lambda} \dfrac{1}{\overline{\phi}(e^{i\lambda})}$

2. $\mathcal{J}^{-1}(P_{L(X:0)}X_n) = \dfrac{1}{\overline{\phi}(e^{i\lambda})} P_{\overline{H^2}}(e^{in\lambda}\overline{\phi}(e^{i\lambda}))$

Proof. (1.) Let $\phi_0(e^{i\lambda}) = \mathcal{J}^{-1}(\nu_0)$ and note that $\mathcal{J}^{-1}(\nu_n) = \mathcal{J}^{-1}(U^n\nu_0) = M_{e^{i\lambda}}^n \mathcal{J}^{-1}(\nu_0) = e^{in\lambda}\phi_0(e^{i\lambda})$, where $M_{e^{i\lambda}}$ denotes the operator of multiplication by $e^{i\lambda}$ on $L^2(\mathbb{T}, F_X)$. It remains to show that $\phi_0(e^{i\lambda}) = \dfrac{1}{\overline{\phi}(e^{i\lambda})}$ $[\sigma]$-a.e. By supposition,

$$X_n = \sum_{k=0}^{\infty} \overline{\phi}(k)\nu_{n-k}. \tag{1.19}$$

Now, applying $\mathcal{J}^{-1}$ to both sides of (1.19), we get

$$e^{in\lambda} = \sum_{k=0}^{\infty} \overline{\phi}(k)e^{i(n-k)\lambda}\phi_0(\lambda) \text{ in } L^2(\mathbb{T}, F_X).$$

Therefore, it follows that

$$1 = \sum_{k=0}^{\infty} \overline{\widehat{\phi}(k)} e^{-ik\lambda} \phi_0(\lambda) \text{ in } L^2(\mathbb{T}, F_X).$$

It then follows from this that

$$\phi(e^{i\lambda}) = \phi(e^{i\lambda})\phi_0(\lambda) \sum_{k=0}^{\infty} \overline{\widehat{\phi}(k)} e^{-ik\lambda} \text{ in } L^2(\mathbb{T}, \sigma).$$

Therefore, there exists a subsequence that converges $[\sigma]$-a.e. That is, there exists a subsequence $(n_j)_{j=0}^{\infty}$ such that

$$\lim_{j \to \infty} \phi(e^{i\lambda})\phi_0(\lambda) \sum_{k=0}^{n_j} \overline{\widehat{\phi}(k)} e^{-ik\lambda} = \phi(e^{i\lambda}) \ [\sigma]\text{-a.e.}$$

Now, since $\phi \neq 0$ $[\sigma]$-a.e., it follows that

$$\lim_{j \to \infty} \phi_0(\lambda) \sum_{k=0}^{n_j} \overline{\widehat{\phi}(k)} e^{-ik\lambda} = 1 \ [\sigma]\text{-a.e.}$$

From this, we see that $\phi_0 \neq 0$ $[\sigma]$-a.e. and therefore, we have

$$\lim_{j \to \infty} \sum_{k=0}^{n_j} \overline{\widehat{\phi}(k)} e^{-ik\lambda} = \frac{1}{\phi_0(\lambda)} \ [\sigma]\text{-a.e.}$$

We now point out that $\left(\sum_{k=0}^{n_j} \overline{\widehat{\phi}(k)} e^{-ik\lambda} \right)_{j=0}^{\infty}$ is a subsequence of $\left(\sum_{k=0}^{n} \overline{\widehat{\phi}(k)} e^{-ik\lambda} \right)_{n=0}^{\infty}$, which we know converges to $\overline{\phi}$ in $L^2(\mathbb{T}, \sigma)$. Therefore, $\left(\sum_{k=0}^{n_j} \overline{\widehat{\phi}(k)} e^{-ik\lambda} \right)_{j=0}^{\infty}$ must also converge to $\overline{\phi}$ in $L^2(\mathbb{T}, \sigma)$, thus having a subsequence that converges to $\overline{\phi}(e^{i\lambda})$ $[\sigma]$-a.e. However, we already have that

$$\lim_{j \to \infty} \sum_{k=0}^{n_j} \overline{\widehat{\phi}(k)} e^{-ik\lambda} = \frac{1}{\phi_0(\lambda)} \ [\sigma]\text{-a.e.}$$

Therefore, we may conclude that $\phi_0(\lambda) = \dfrac{1}{\overline{\phi}(e^{i\lambda})}$ $[\sigma]$-a.e.

(2.) By supposition,

$$X_n = \sum_{k=0}^{\infty} \overline{\widehat{\phi}(k)} \nu_{n-k}. \tag{1.20}$$

and $L(X:n) = L(\nu:n)$. Therefore, we see that

$$P_{L(X:0)}X_n = P_{L(\nu:0)}X_n = \sum_{k=n}^{\infty} \overline{\hat{\phi}}(k)\nu_{n-k}.$$

Now applying $\mathcal{J}^{-1}$, we get that

$$\begin{aligned}
\mathcal{J}^{-1}(P_{L(X:0)}X_n) &= \sum_{k=n}^{\infty} \overline{\hat{\phi}}(k)e^{i(n-k)\lambda}\frac{1}{\overline{\phi}(e^{i\lambda})} \\
&= \frac{1}{\overline{\phi}(e^{i\lambda})} \sum_{k=0}^{\infty} \overline{\hat{\phi}}(k+n)e^{-ik\lambda} \\
&= \frac{1}{\overline{\phi}(e^{i\lambda})} P_{\overline{H^2}}(e^{in\lambda}\overline{\phi}(e^{i\lambda})).
\end{aligned}$$

$\square$

In practice, we do not know the innovation processes as above, but observe X_n. The following theorems of Kolmogorov examine the representation of ν_n in terms of X_n, $n \in \mathbb{Z}$.

Theorem 1.14.2. *Let X_n, $n \in \mathbb{Z}$ be a weakly stationary regular sequence with maximal factor ϕ and innovation process ν_n, $n \in \mathbb{Z}$. Then, if*

$$\nu_n = \sum_{k=0}^{\infty} d_k X_{n-k}, \text{ with } \sum_{k=0}^{\infty} |d_k|^2 < \infty$$

then $\dfrac{1}{\phi} \in L^2(\mathbb{T}, \sigma)$ *and* $d_k = \overline{\left(\dfrac{1}{\phi}\right)}(k)$.

Proof. By supposition, we have that

$$\nu_n = \sum_{k=0}^{\infty} d_k X_{n-k} \text{ for all } n \in \mathbb{Z}, \tag{1.21}$$

with $\displaystyle\sum_{k=0}^{\infty} |d_k|^2 < \infty$. Now, applying $\mathcal{J}^{-1}$ to both sides of (1.21) and using Theorem 1.14.1, we get that

$$\frac{e^{in\lambda}}{\overline{\phi}(e^{i\lambda})} = \sum_{k=0}^{\infty} d_k e^{i(n-k)\lambda} \text{ in } L^2(\mathbb{T}, F_X).$$

It follows from this that

$$\frac{1}{\overline{\phi}(e^{i\lambda})} = \sum_{k=0}^{\infty} d_k e^{-ik\lambda} \text{ in } L^2(\mathbb{T}, F_X).$$

Therefore, there exists a subsequence that converges $[F_X]$-a.e. That is, there exists a subsequence $(n_j)_{j=0}^{\infty}$ such that

$$\lim_{j\to\infty} \sum_{k=0}^{n_j} d_k e^{-ik\lambda} = \frac{1}{\overline{\phi}(e^{i\lambda})} \ [F_X]\text{-a.e.}$$

Now, since $\phi \neq 0$ $[\sigma]$-a.e., it follows that the above limit converges $[\sigma]$-a.e. Using the fact that $\sum_{k=0}^{\infty} |d_k|^2 < \infty$, we may conclude that

$$\lim_{j\to\infty} \sum_{k=0}^{n_j} d_k e^{-ik\lambda} = \psi(e^{i\lambda}) \text{ in } L^2(\mathbb{T}, \sigma) \text{ for some } \psi \in L^2(\mathbb{T}, \sigma).$$

It follows that a subsequence then converges $[\sigma]$-a.e. to $\psi(e^{i\lambda})$. Then recalling that

$$\lim_{j\to\infty} \sum_{k=0}^{n_j} d_k e^{-ik\lambda} = \frac{1}{\overline{\phi}(e^{i\lambda})} \ [\sigma]\text{-a.e.,}$$

we may conclude that $\dfrac{1}{\overline{\phi}(e^{i\lambda})} = \psi(e^{i\lambda})$ $[\sigma]$-a.e. and hence $\dfrac{1}{\phi} \in L^2(\mathbb{T}, \sigma)$. We finish the proof by pointing out that since $\left(\sum_{k=0}^{n} d_k e^{-ik\lambda}\right)_{n=0}^{\infty}$ converges in $L^2(\mathbb{T}, \sigma)$ and $\left(\sum_{k=0}^{n_j} d_k e^{-ik\lambda}\right)_{j=0}^{\infty}$ is a subsequence that converges to $\dfrac{1}{\overline{\phi}(e^{i\lambda})}$ in $L^2(\mathbb{T}, \sigma)$, that $\left(\sum_{k=0}^{n} d_k e^{-ik\lambda}\right)_{n=0}^{\infty}$ must also converge to $\dfrac{1}{\overline{\phi}(e^{i\lambda})}$ in $L^2(\mathbb{T}, \sigma)$. It then follows from the uniqueness of the Fourier coefficients that $d_k = \overline{\left(\dfrac{1}{\phi}\right)}(k)$. $\qquad\square$

Theorem 1.14.3. *Let X_n, $n \in \mathbb{Z}$ be a weakly stationary regular sequence with maximal factor ϕ and innovation process ν_n, $n \in \mathbb{Z}$. Then, if $\dfrac{1}{\phi} \in L^2(\mathbb{T}, \sigma)$ and $\phi \in L^\infty(\mathbb{T}, \sigma)$ then*

$$\nu_n = \sum_{k=0}^{\infty} \overline{\left(\frac{1}{\phi}\right)}(k) X_{n-k}.$$

Proof. Since ϕ is maximal and $\dfrac{1}{\phi} \in L^2(\mathbb{T}, \sigma)$, it follows from H^p-theory that

$$\sum_{k=0}^{\infty} \overline{\left(\frac{1}{\phi}\right)}(k) e^{-ik\lambda} = \frac{1}{\overline{\phi}(e^{i\lambda})} \text{ in } L^2(\mathbb{T}, \sigma).$$

The following inequality shows that

$$\sum_{k=0}^{\infty} \overline{\left(\frac{\widehat{1}}{\phi}\right)}(k)e^{-ik\lambda} = \frac{1}{\overline{\phi(e^{i\lambda})}} \quad \text{in } L^2(\mathbb{T}, F_X).$$

$$\int_{[-\pi,\pi)} \left| \sum_{k=0}^{n} \overline{\left(\frac{\widehat{1}}{\phi}\right)}(k)e^{-ik\lambda} - \frac{1}{\overline{\phi(e^{i\lambda})}} \right|^2 |\phi(e^{i\lambda})|^2 \, d\sigma(e^{i\lambda}) \leq$$

$$\||\phi|^2\|_{L^\infty} \int_{[-\pi,\pi)} \left| \sum_{k=0}^{n} \overline{\left(\frac{\widehat{1}}{\phi}\right)}(k)e^{-ik\lambda} - \frac{1}{\overline{\phi(e^{i\lambda})}} \right|^2 d\sigma(e^{i\lambda}).$$

Now, we may apply $\mathcal{J}$ to both sides and get

$$\nu_0 = \sum_{k=0}^{\infty} \overline{\left(\frac{\widehat{1}}{\phi}\right)}(k) X_{-k}.$$

Therefore,

$$\nu_n = U^n \nu_0 = \sum_{k=0}^{\infty} \overline{\left(\frac{\widehat{1}}{\phi}\right)}(k) U^n X_{-k} = \sum_{k=0}^{\infty} \overline{\left(\frac{\widehat{1}}{\phi}\right)}(k) X_{n-k},$$

as desired.   □

1.15   The Szegö-Krein-Kolmogorov Theorem

Let X_n, $n \in \mathbb{Z}$ be a weakly stationary sequence with spectral measure F_X. We define the one-step mean square error to be

$$\varepsilon^2(F_X) = \inf_{c_j, m} \left\| X_0 - \sum_{j=1}^{m} c_j X_{-j} \right\|_{L(X)}^2$$

$$= \inf_{c_j, m} \int_{[-\pi,\pi)} \left| 1 - \sum_{j=1}^{m} c_j e^{-ij\lambda} \right|^2 dF_X(e^{i\lambda}).$$

We will prove the following theorem.

Theorem 1.15.1. $\varepsilon^2(F_X) = \exp\left\{ \int_{[-\pi,\pi)} \log f_X(e^{i\lambda}) \, d\sigma(e^{i\lambda}) \right\}$, where f_X is the density of the absolutely continuous part of F_X with respect to σ.

Before we prove this theorem, we state a corollary that follows immediately from this theorem.

Corollary 1.15.1. X_n, $n \in \mathbb{Z}$ *is singular if and only if* $\int_{[-\pi,\pi)} \log f_X(e^{i\lambda}) \, d\sigma(e^{i\lambda}) = -\infty$.

Now, in the direction to prove our theorem, we will show that

$$\varepsilon^2(F_X) = \inf_{c_j, m} \int_{[-\pi,\pi)} \left| 1 - \sum_{j=1}^{m} c_j e^{-ij\lambda} \right|^2 f_X(e^{i\lambda}) \, d\sigma(e^{i\lambda}). \qquad (1.22)$$

First, if X_n, $n \in \mathbb{Z}$ is not singular, then by the Wold Decomposition (Theorem 1.11.1), we have that

$$\left\| X_0 - \sum_{j=1}^{m} c_j X_{-j} \right\|^2 = \left\| X_0^r - \sum_{j=1}^{m} c_j X_{-j}^r \right\|^2 + \left\| X_0^s - \sum_{j=1}^{m} c_j X_{-j}^s \right\|^2,$$

where X_n^r is regular and X_n^s is singular for all n in $\mathbb{Z}$. By the definition of singular, we get that $\left\| X_0^s - \sum_{j=1}^{m} c_j X_{-j}^s \right\| = 0$, for all c_j and m. From these observations and the Concordance Theorem (Theorem 1.13.1), Equation (1.22) follows. However, if X_n, $n \in \mathbb{Z}$ is singular, we must take a different point of view. Let $H_0^2(\mathbb{T}, F_X)$ denote the closure in $L^2(\mathbb{T}^2, F_X)$ of the span of $\{e^{in\lambda} : n \geq 1\}$. Let $1 + H_0^2(\mathbb{T}, F_X)$ denote the set of all elements of the form $1 + f$, where f is a member of $H_0^2(\mathbb{T}, F_X)$. $1 + H_0^2(\mathbb{T}, F_X)$ forms a nonempty, closed, convex subset of $L^2(\mathbb{T}, F_X)$. As such, it contains a unique element of smallest norm. If this element is zero, then we say that prediction is perfect. Otherwise, we get a nonzero element of minimal norm of the form $1 + H$, where H is in $H_0^2(\mathbb{T}, F_X)$.

For any complex number c and $m \geq 1$,

$$1 + H(e^{i\lambda}) + c e^{-im\lambda}$$

belongs to $1 + H_0^2(\mathbb{T}, F_X)$. Since $1 + H$ is the unique element of minimal norm in $1 + H_0^2(\mathbb{T}, F_X)$, we know that

$$\int_{[-\pi,\pi)} \left| 1 + H(e^{i\lambda}) + c e^{-im\lambda} \right|^2 \, dF_X(e^{i\lambda})$$

has a unique minimum at $c = 0$. A straightforward calculation then shows that for every $m \geq 1$,

$$\int_{[-\pi,\pi)} \left(1 + H(e^{i\lambda}) \right) e^{im\lambda} \, dF_X(e^{i\lambda}) = 0. \qquad (1.23)$$

Since S is closed under addition, we get a second orthogonality relation. For each complex number c and each $m \geq 1$, the function

$$\left(1 + H(e^{i\lambda})\right)\left(1 + c\,e^{-im\lambda}\right)$$

belongs to $1 + H_0^2(\mathbb{T}, F_X)$, and its norm is minimized at $c = 0$. Therefore,

$$\int_{[-\pi,\pi)} \left|1 + H(e^{i\lambda})\right|^2 e^{im\lambda} \, dF_X(e^{i\lambda}) = 0, \qquad (1.24)$$

for all $m \geq 1$. By taking the complex conjugate of (1.24), we get that the same equation holds for all $m \leq -1$. That is, the Fourier-Stieltjes coefficients of the measure $\left|1 + H(e^{i\lambda})\right|^2 dF_X(e^{i\lambda})$ all vanish except the central one. Therefore, this measure is a multiple of σ, normalized Lebesgue measure. It follows that $1 + H$ must vanish almost everywhere with respect to the singular component of dF_X, and (1.23) can be written as,

$$\int_{[-\pi,\pi)} \left(1 + H(e^{i\lambda})\right) e^{im\lambda} \, dF_{Xa}(e^{i\lambda}) = 0, \qquad m \geq 1, \qquad (1.25)$$

where F_{Xa} is the absolutely continuous part of F_X with respect to σ.

Next, we observe that (1.23) characterizes the minimal element in $1 + H_0^2(\mathbb{T}, F_X)$. To see this, suppose that (1.23) holds, but $1 + G$ is the minimal element. Then,

$$\int_{[-\pi,\pi)} \left|1 + H(e^{i\lambda}) + c\left(G(e^{i\lambda}) - H(e^{i\lambda})\right)\right|^2 dF_X(e^{i\lambda})$$

$$= \int_{[-\pi,\pi)} \left|1 + H(e^{i\lambda})\right|^2 dF_X(e^{i\lambda})$$

$$+ \ |c|^2 \int_{[-\pi,\pi)} \left|G(e^{i\lambda}) - H(e^{i\lambda})\right|^2 dF_X(e^{i\lambda}),$$

for every complex number c. The right-hand form of the expression is clearly smallest when $c = 0$, and the left-hand form of the expression is certainly smallest when $c = 1$ since $1 + G$ is the minimal element. From these observations and the fact that the minimal element is unique, we conclude that $G = H$.

If $\varepsilon^2(F_X)$ is positive, we have established the fact that it is equal to

$$\int_{[-\pi,\pi)} \left|1 + H(e^{i\lambda})\right|^2 dF_X(e^{i\lambda}),$$

where $1 + H$ belongs to $1 + H_0^2(\mathbb{T}, F_X)$ and vanishes almost everywhere with respect to the singular component of dF_X. Hence, (1.25) holds. Moreover, $1 + H$ belongs to the convex set $1 + H_0^2(\mathbb{T}, f_X \, d\sigma)$. Note the closure is now with respect to $f_X \, d\sigma$ rather than F_X. Furthermore, (1.25) implies that $1 + H$ is the minimal function relative to this measure and so

$$\inf_{c_j, m} \int_{[-\pi, \pi)} \left| 1 - \sum_{j=1}^{m} c_j e^{-ij\lambda} \right|^2 f_X(e^{i\lambda}) \, d\sigma(e^{i\lambda})$$

$$= \int_{[-\pi, \pi)} \left| 1 + H(e^{i\lambda}) \right|^2 f_X(e^{i\lambda}) \, d\sigma(e^{i\lambda})$$

$$= \int_{[-\pi, \pi)} \left| 1 + H(e^{i\lambda}) \right|^2 \, dF_X(e^{i\lambda}).$$

Therefore, it will suffice to prove the following theorem.

Theorem 1.15.2. $\displaystyle \inf_{c_j, m} \int_{[-\pi, \pi)} \left| 1 - \sum_{j=1}^{m} c_j e^{-ij\lambda} \right|^2 f_X(e^{i\lambda}) \, d\sigma(e^{i\lambda})$

$$= \exp \left\{ \int_{[-\pi, \pi)} \log f_X(e^{i\lambda}) \, d\sigma(e^{i\lambda}) \right\}.$$

We note that it is also enough to prove Theorem 1.15.2 for the case when $\varepsilon^2(F_X)$ is zero. To prove Theorem 1.15.2, we use two lemmas. The proofs are due to Helson and Lowdenslager, see [10]. Before we state the lemmas, we recall that $\hat{\psi}(0) := \int_{[-\pi, \pi)} \psi(e^{i\lambda}) \, d\sigma(e^{i\lambda})$.

Lemma 1.15.1. *If f is nonnegative $[\sigma]$-a.e. with f in $L^1(\mathbb{T}, \sigma)$, then*

$$\exp \left\{ \int_{[-\pi, \pi)} \log f(e^{i\lambda}) \, d\sigma(e^{i\lambda}) \right\} = \inf_{\psi} \int_{[-\pi, \pi)} e^{\psi(e^{i\lambda})} f(e^{i\lambda}) \, d\sigma(e^{i\lambda}), \text{ where the}$$

infimum is taken over all real-valued functions ψ in $L^1(\mathbb{T}, \sigma)$ such that $\hat{\psi}(0) = 0$.

Proof. First, suppose that $\log f$ in $L^1(\mathbb{T}, \sigma)$. Then, we have

$$\exp \left\{ \int_{[-\pi, \pi)} \log f(e^{i\lambda}) \, d\sigma(e^{i\lambda}) \right\}$$

$$= \exp \left\{ \int_{[-\pi, \pi)} \log e^{\psi(e^{i\lambda})} \, d\sigma(e^{i\lambda}) + \int_{[-\pi, \pi)} \log f(e^{i\lambda}) \, d\sigma(e^{i\lambda}) \right\}$$

$$= \exp \left\{ \int_{[-\pi, \pi)} \log \left(e^{\psi(e^{i\lambda})} f(e^{i\lambda}) \right) \, d\sigma(e^{i\lambda}) \right\}$$

$$\leq \int_{[-\pi, \pi)} e^{\psi(e^{i\lambda})} f(e^{i\lambda}) \, d\sigma(e^{i\lambda}).$$

The last inequality follows from Jensen's Inequality. Therefore,

$$\exp\left\{\int_{[-\pi,\pi)} \log f(e^{i\lambda})\,d\sigma(e^{i\lambda})\right\} \le \inf_{\psi} \int_{[-\pi,\pi)} e^{\psi(e^{i\lambda})} f(e^{i\lambda})\,d\sigma(e^{i\lambda}),$$

where the infimum is taken over all real-valued functions ψ in $L^1(\mathbb{T},\sigma)$ such that $\hat{\psi}(0) = 0$. To see that equality holds, let $\psi(e^{i\lambda}) = \int_{[-\pi,\pi)} \log f(e^{i\lambda})\,d\sigma(e^{i\lambda}) - \log f(e^{i\lambda})$. Then ψ satisfies all of the conditions of the lemma and

$$\exp\left\{\int_{[-\pi,\pi)} \log f(e^{i\lambda})\,d\sigma(e^{i\lambda})\right\} = \int_{[-\pi,\pi)} e^{\psi(e^{i\lambda})} f(\lambda)\,d\sigma(e^{i\lambda}).$$

Therefore,

$$\exp\left\{\int_{[-\pi,\pi)} \log f(e^{i\lambda})\,d\sigma(e^{i\lambda})\right\} = \inf_{\psi} \int_{[-\pi,\pi)} e^{\psi(e^{i\lambda})} f(e^{i\lambda})\,d\sigma(e^{i\lambda}).$$

Now, suppose that $\log f$ is not in $L^1(\mathbb{T},\sigma)$. Then, $\log(f+\epsilon)$ is in $L^1(\mathbb{T},\sigma)$, for every $\epsilon > 0$. Therefore, it follows from above that

$$
\begin{aligned}
\exp\left\{\int_{[-\pi,\pi)} \log(f(e^{i\lambda})+\epsilon)\,d\sigma(e^{i\lambda})\right\} &= \inf_{\psi} \int_{[-\pi,\pi)} e^{\psi(e^{i\lambda})}(f(e^{i\lambda})+\epsilon)\,d\sigma(e^{i\lambda}). \\
&\ge \inf_{\psi} \int_{[-\pi,\pi)} e^{\psi(e^{i\lambda})} f(e^{i\lambda})\,d\sigma(e^{i\lambda}) \ge 0.
\end{aligned}
$$

By the Monotone Limit Theorem, the left-hand side of the above equation converges to

$$\exp\left\{\int_{[-\pi,\pi)} \log f(e^{i\lambda})\,d\sigma(e^{i\lambda})\right\} = 0,$$

as ϵ tends to zero. Therefore, in this case as well,

$$\exp\left\{\int_{[-\pi,\pi)} \log f(e^{i\lambda})\,d\sigma(e^{i\lambda})\right\} = \inf_{\psi} \int_{[-\pi,\pi)} e^{\psi(e^{i\lambda})} f(e^{i\lambda})\,d\sigma(e^{i\lambda}).$$

$\square$

Lemma 1.15.2. *If f is nonnegative $[\sigma]$-a.e. with f in $L^1(\mathbb{T},\sigma)$, then*

$$\exp\left\{\int_{[-\pi,\pi)} \log f(e^{i\lambda})\,d\sigma(e^{i\lambda})\right\} = \inf_{\psi} \int_{[-\pi,\pi)} e^{\psi(e^{i\lambda})} f(e^{i\lambda})\,d\sigma(e^{i\lambda}), \text{ where the}$$

infimum is taken over all real-valued trigonometric polynomials ψ such that $\hat{\psi}(0) = 0$.

Proof. We will assume that $\log f$ in $L^1(\mathbb{T}, \sigma)$, since we may use a limiting argument, as in Lemma 1.15.1, for the general case. It follows from the proof of Lemma 1.15.1 that

$$\exp\left\{\int_{[-\pi,\pi)} \log f(e^{i\lambda}) \, d\sigma(e^{i\lambda})\right\} \leq \inf_{\psi} \int_{[-\pi,\pi)} e^{\psi(e^{i\lambda})} f(e^{i\lambda}) \, d\sigma(e^{i\lambda}),$$

where the infimum is taken over all real-valued trigonometric polynomials ψ such that $\hat{\psi}(0) = 0$. For the other direction, we assume that

$$\int_{[-\pi,\pi)} \log f(e^{i\lambda}) \, d\sigma(e^{i\lambda}) = 0.$$

If this were not the case, we can divide f by $k = \exp\left\{\int_{[-\pi,\pi)} \log f(e^{i\lambda}) \, d\sigma(e^{i\lambda})\right\}$.

Then,

$$\int_{[-\pi,\pi)} \log \frac{f(e^{i\lambda})}{k} \, d\sigma(e^{i\lambda}) = 0.$$

We could then work out the details of the proof with f/k in place of f. Since,
$\int_{[-\pi,\pi)} \log f(e^{i\lambda}) \, d\sigma(e^{i\lambda}) = 0.$ It remains to show that

$$\inf_{\psi} \int_{[-\pi,\pi)} e^{\psi(e^{i\lambda})} f(e^{i\lambda}) \, d\sigma(e^{i\lambda}) \leq 1,$$

where the infimum is taken over all real-valued trigonometric polynomials ψ such that $\hat{\psi}(0) = 0$. Since every bounded function ψ is boundedly the limit of Fejér means of its Fourier series, with each approximating function P a trigonometric polynomial that is real-valued if ψ is real-valued, and $\hat{P}(0) = 0$ if $\hat{\psi}(0) = 0$, we will show this final inequality for bounded real-valued functions ψ and then our desired result follows.

Let $\left(u_n(e^{i\lambda})\right)_{n=1}^{\infty}$ be a sequence of nonnegative bounded functions that increase pointwise to $\log^+ f(e^{i\lambda})$, and let $\left(v_n(e^{i\lambda})\right)_{n=1}^{\infty}$ be a sequence of nonnegative bounded functions that increase pointwise to $\log^- f(e^{i\lambda})$. Then, by the monotone limit theorem,

$$\lim_{n \to \infty} \int_{[-\pi,\pi)} u_n(e^{i\lambda}) \, d\sigma(e^{i\lambda}) = \int_{[-\pi,\pi)} \log^+ f(e^{i\lambda}) \, d\sigma(e^{i\lambda})$$

$$= \int_{[-\pi,\pi)} \log^- f(e^{i\lambda}) \, d\sigma(e^{i\lambda}) = \lim_{n \to \infty} \int_{[-\pi,\pi)} v_n(e^{i\lambda}) \, d\sigma(e^{i\lambda}).$$

It then follows that for each n, there exists an m such that

$$\int_{[-\pi,\pi)} u_n(e^{i\lambda}) \, d\sigma(e^{i\lambda}) \leq \int_{[-\pi,\pi)} v_m(e^{i\lambda}) \, d\sigma(e^{i\lambda}).$$

We can multiply v_m by a positive constant less than or equal to one, and rename the function v_n, so that

$$\int_{[-\pi,\pi)} u_n(e^{i\lambda})\, d\sigma(e^{i\lambda}) = \int_{[-\pi,\pi)} v_n(e^{i\lambda})\, d\sigma(e^{i\lambda}).$$

This new sequence $(v_n(e^{i\lambda}))_{n=1}^{\infty}$ still converges pointwise to $\log^- f(e^{i\lambda})$, although the convergence may no longer be monotonic. It follows from the construction that

$$0 \le e^{(\log^+ f - u_n) - (\log^- f - v_n)} \le \max\{1, f\}.$$

Therefore, we can apply the Lebesgue dominated convergence theorem to get

$$\lim_{n \to \infty} \int_{[-\pi,\pi)} e^{(\log^+ f(e^{i\lambda}) - u_n(e^{i\lambda})) - (\log^- f(e^{i\lambda}) - v_n(e^{i\lambda}))}\, d\sigma(e^{i\lambda}) = 1.$$

Since

$$\int_{[-\pi,\pi)} e^{(\log^+ f(e^{i\lambda}) - u_n(e^{i\lambda})) - (\log^- f(e^{i\lambda}) - v_n(e^{i\lambda}))}\, d\sigma(e^{i\lambda}) =$$

$$\int_{[-\pi,\pi)} e^{v_n(e^{i\lambda}) - u_n(e^{i\lambda})} f(e^{i\lambda})\, d\sigma(e^{i\lambda}),$$

$v_n - u_n$ is a real-valued bounded function with $\widehat{(v_n - u_n)}(0) = 0$, it follows that

$$\inf_{\psi} \int_{[-\pi,\pi)} e^{\psi(e^{i\lambda})} f(e^{i\lambda})\, d\sigma(e^{i\lambda}) \le 1,$$

where the infimum is taken over all real-valued bounded functions ψ such that $\hat{\psi}(0) = 0$.   □

We will use this lemma to prove Theorem 1.15.2. The proof is due to Helson and Lowdenslager, see [10].

Proof of Theorem 1.15.2. First, real-valued trigonometric polynomials satisfying the conditions of Lemma 1.15.2 can be represented in the form $P(e^{i\lambda}) + \overline{P(e^{i\lambda})}$, where $P(e^{i\lambda}) = \sum_{n=1}^{N} a_n e^{in\lambda}$. Therefore, we may write

$$e^{\psi(e^{i\lambda})} = e^{P(e^{i\lambda}) + \overline{P(e^{i\lambda})}} = \left| e^{P(e^{i\lambda})} \right|^2.$$

If we define $Q(z) = e^{P(z)} - 1$, then $Q(0) = 0$ and $Q(z)$ is an entire function. We may now write

$$e^{\psi(e^{i\lambda})} = \left| 1 + Q(e^{i\lambda}) \right|^2.$$

Hence, it follows from our above lemma that

$$\exp\left\{ \int_{[-\pi,\pi)} \log f(e^{i\lambda})\, d\sigma(e^{i\lambda}) \right\} \ge \inf_{Q} \int_{[-\pi,\pi)} \left| 1 + Q(e^{i\lambda}) \right|^2 f(e^{i\lambda})\, d\sigma(e^{i\lambda}),$$

where the infimum is taken over all entire functions Q with $Q(0) = 0$. Since entire functions can be uniformly approximated by polynomials on compact subsets, it follows that we can replace entire functions by polynomials. So, we have

$$\exp\left\{\int_{[-\pi,\pi)} \log f(e^{i\lambda})\, d\sigma(e^{i\lambda})\right\} \geq \inf_P \int_{[-\pi,\pi)} \left|1 + P(e^{i\lambda})\right|^2 f(e^{i\lambda})\, d\sigma(e^{i\lambda}),$$

where the infimum is taken over all polynomial functions P with $P(0) = 0$. It remains to show that this inequality is actually an equality. To do this, we start by taking $f = |1 + Q|^2$, where Q is a polynomial with $Q(0) = 0$. We then have

$$\exp\left\{\int_{[-\pi,\pi)} \log \left|1 + Q(e^{i\lambda})\right|^2\, d\sigma(e^{i\lambda})\right\}$$

$$\geq \inf_P \int_{[-\pi,\pi)} \left|1 + P(e^{i\lambda})\right|^2 \left|1 + Q(e^{i\lambda})\right|^2\, d\sigma(e^{i\lambda})$$

$$= \inf_P \int_{[-\pi,\pi)} \left|1 + P(e^{i\lambda}) + Q(e^{i\lambda}) + P(e^{i\lambda})Q(e^{i\lambda})\right|^2\, d\sigma(e^{i\lambda}) \geq 1,$$

where the infimum is taken over all polynomial functions P with $P(0) = 0$. Hence, $\log |1 + Q|^2$ is in $L^1(\mathbb{T}, \sigma)$ and

$$\int_{[-\pi,\pi)} \log \left|1 + Q(e^{i\lambda})\right|^2\, d\sigma(e^{i\lambda}) \geq 0.$$

Let $k = \exp\left\{\int_{[-\pi,\pi)} \log \left|1 + Q(e^{i\lambda})\right|^2\, d\sigma(e^{i\lambda})\right\}$. Then, $k \geq 1$ and if we set $\varphi = \log |1 + Q|^2 - \log k$, we see that φ is a real-valued function in $L^1(\mathbb{T}, \sigma)$, with $\int_{[-\pi,\pi)} \varphi(e^{i\lambda})\, d\sigma(e^{i\lambda}) = 0$. Therefore, $|1 + Q|^2 = ke^{\varphi}$ and if we go back to our original f, we have

$$\int_{[-\pi,\pi)} \left|1 + Q(e^{i\lambda})\right|^2 f(e^{i\lambda})\, d\sigma(e^{i\lambda}) = k \int_{[-\pi,\pi)} e^{\varphi(e^{i\lambda})} f(e^{i\lambda})\, d\sigma(e^{i\lambda})$$

$$\geq \int_{[-\pi,\pi)} e^{\varphi(e^{i\lambda})} f(e^{i\lambda})\, d\sigma(e^{i\lambda})$$

$$\geq \inf_\psi \int_{[-\pi,\pi)} e^{\psi(e^{i\lambda})} f(e^{i\lambda})\, d\sigma(e^{i\lambda})$$

$$= \exp\left\{\int_{[-\pi,\pi)} \log f(e^{i\lambda})\, d\sigma(e^{i\lambda})\right\}.$$

Now, taking the infimum over all polynomials Q with $Q(0) = 0$, we get the other opposite inequality and hence,

$$\exp\left\{\int_{[-\pi,\pi)} \log f(e^{i\lambda})\, d\sigma(e^{i\lambda})\right\} = \inf_{P} \int_{[-\pi,\pi)} \left|1 + P(e^{i\lambda})\right|^2 f(e^{i\lambda})\, d\sigma(e^{i\lambda}),$$

where the infimum is taken over all polynomial functions P with $P(0) = 0$. Finally, since complex conjugation does not affect the modulus, we have

$$\exp\left\{\int_{[-\pi,\pi)} \log f(e^{i\lambda})\, d\sigma(e^{i\lambda})\right\} = \inf_{P} \int_{[-\pi,\pi)} \left|1 + P(e^{-i\lambda})\right|^2 f(e^{i\lambda})\, d\sigma(e^{i\lambda}),$$

where the infimum is taken over all polynomial functions P with $P(0) = 0$. More explicitly, we have

$$\exp\left\{\int_{[-\pi,\pi)} \log f(e^{i\lambda})\, d\sigma(e^{i\lambda})\right\} = \inf_{c_j, m} \int_{[-\pi,\pi)} \left|1 - \sum_{j=1}^{m} c_j e^{-ij\lambda}\right|^2 f(e^{i\lambda})\, d\sigma(e^{i\lambda}).$$

Now, since f_X satisfies the conditions of our lemma, we may conclude that

$$\exp\left\{\int_{[-\pi,\pi)} \log f_X(e^{i\lambda})\, d\sigma(e^{i\lambda})\right\}$$

$$= \inf_{c_j, m} \int_{[-\pi,\pi)} \left|1 - \sum_{j=1}^{m} c_j e^{-ij\lambda}\right|^2 f_X(e^{i\lambda})\, d\sigma(e^{i\lambda}),$$

as desired.   □

We finish this discussion with one final comment regarding the one-step mean square error. If X_n, $n \in \mathbb{Z}$ is not singular, then we have seen that the error is determined only by the regular part of the sequence and every regular sequence has a maximal factor, say ϕ and innovation process, say ν_n, $n \in \mathbb{Z}$ with $L(X : n) = L(\nu : n)$ for all $n \in \mathbb{Z}$, it follows, in the language of operator theory, that

$$\varepsilon^2(F_X) = E|X_0 - P_{L(X:-1)} X_0|^2 = E|X_0 - P_{L(\nu:-1)} X_0|^2 = |\widehat{\phi}(0)|^2.$$

We leave this section with one final corollary.

Corollary 1.15.2. *A weakly stationary sequence X_n, $n \in \mathbb{Z}$ is regular if and only if its spectral measure F_X is absolutely continuous with respect to normalized Lebesgue measure and has density f_X that is positive $[\sigma]$-a.e. satisfying*

$$\int_{[-\pi,\pi)} \log f_X(e^{i\lambda})\, d\sigma(e^{i\lambda}) > -\infty.$$

Proof. Suppose X_n, $n \in \mathbb{Z}$ is regular. By Theorem 1.12.2, we get that F_X is absolutely continuous with respect to normalized Lebesgue measure with density $f_X = |\varphi|^2$, where $\varphi \in H^2(\mathbb{T})$. It is well known, see the second corollary on page 52 of [13], that such a function is positive $[\sigma]$-a.e. and by the first theorem on page 53 of [13], $\displaystyle\int_{[-\pi,\pi)} \log f_X(e^{i\lambda}) \, d\sigma(e^{i\lambda}) > -\infty$. For the other direction, we use our supposition and the first theorem on page 53 of [13] to conclude that $f_X = |\varphi|^2$, where $\varphi \in H^2(\mathbb{T})$. Then, by Theorem 1.12.3, X_n, $n \in \mathbb{Z}$ is regular. □

1.16 Remarks and Related Literature

The presentation in this chapter is based on the references ([19], [23], and [45]). For additional reading, one might consider the books [Y] and [MR]. These books also cover stationary sequences. The first reference, [Y], is a textbook that covers some parts of this chapter, and the second reference, [MR], studies, in addition, some special examples which arise in statistical application.

[Y] A. M. Yaglom, *An Introduction to the Theory of Stationary Random Functions* (Revised English Edition. Translated by Richard A. Silverman), Prentice Hall, Inc. Engelwood Cliffs, N.J., 1962.

[MR] Murray Rosenblatt, *Stationary Sequences and Random Fields*, Biskhauser, Boston, 1985.

2

Weakly Stationary Random Fields

2.1 Preliminaries

We call a doubly indexed sequence of random variables, $X_{m,n}$, $(m,n) \in \mathbb{Z}^2$, on $(\Omega, \mathcal{F}, P)$ a random field. We say a random field $X_{m,n}$, $(m,n) \in \mathbb{Z}^2$ is a second order random field, if $E|X_{m,n}|^2 < \infty$, $(m,n) \in \mathbb{Z}^2$. That is, $X_{m,n} \in L^2(\Omega, \mathcal{F}, P)$, $(m,n) \in \mathbb{Z}^2$. A second order random field $X_{m,n}$, $(m,n) \in \mathbb{Z}^2$ is called a weakly stationary random field, if

1. $EX_{m,n} = C$, $(m,n) \in \mathbb{Z}^2$, and

2. $\mathrm{cov}(X_{m,n}, X_{m',n'}) := E(X_{m,n} - EX_{m,n})\overline{(X_{m',n'} - EX_{m',n'})}$, $(m,n),(m',n') \in \mathbb{Z}^2$ depends only on $m - m'$ and $n - n'$. That is,

$$\mathrm{cov}(X_{m+k,n+l}, X_{k,l}) = \mathrm{cov}(X_{m,n}, X_{0,0}), \qquad ((m,n),(k,l) \in \mathbb{Z}^2).$$

For the sake of simplicity, we assume that $EX_{m,n} = 0$, $(m,n) \in \mathbb{Z}^2$. We point out that, under this assumption, the $\mathrm{cov}(X_{m,n}, X_{m',n'})$ is just the inner product of $X_{m,n}$ and $X_{m',n'}$ in $L^2(\Omega, \mathcal{F}, P)$. For a weakly stationary random field $X_{m,n}$, $(m,n) \in \mathbb{Z}^2$, we define the covariance function to be

$$r_X(m,n) = \mathrm{cov}(X_{m,n}, X_{0,0}), \qquad ((m,n) \in \mathbb{Z}^2).$$

Given a weakly stationary random field $X_{m,n}$, $(m,n) \in \mathbb{Z}^2$, let $L(X)$ denote the (closed) subspace of $L^2(\Omega, \mathcal{F}, P)$ generated by $X_{m,n}$, $(m,n) \in \mathbb{Z}^2$. That is, let

$$L(X) = \overline{\mathrm{span}}\{X_{m,n} : (m,n) \in \mathbb{Z}^2\}.$$

We now show the existence of two unitary operators, U_1 and U_2, defined on $L(X)$ with the property that $U_1 X_{m,n} = X_{m+1,n}$ and $U_2 X_{m,n} = X_{m,n+1}$ for all $(m,n) \in \mathbb{Z}^2$. We will then use these operators and a spectral-type theorem, found in the appendix, for a commutative B^*-algebra generated by two commuting unitary operators to give the spectral representation theorem for weakly stationary random fields.

We now show the existence of U_1. The existence of U_2 is shown in an analogous way. For each $(m,n) \in \mathbb{Z}^2$, define $U_1 X_{m,n} = X_{m+1,n}$. Extending by linearity, we get for a finite linear combination that

$$U_1 \left(\sum_{k,l} a_{k,l} X_{k,l} \right) = \sum_{k,l} a_{k,l} X_{k+1,l}.$$

To see that U_1 is well defined, note that if

$$\sum_{k,l} a_{k,l} X_{k,l} = \sum_{k,l} b_{k,l} X_{k,l}$$

then

$$\left\| U_1 \left(\sum_{k,l} a_{k,l} X_{k,l} \right) - U_1 \left(\sum_{k,l} b_{k,l} X_{k,l} \right) \right\|_{L(X)}^2$$

$$= \left\| \sum_{k,l} a_{k,l} X_{k+1,l} - \sum_{k,l} b_{k,l} X_{k+1,l} \right\|_{L(X)}^2$$

$$= \sum_{k,l} \sum_{m,n} (a_{k,l} - b_{k,l}) \overline{(a_{m,n} - b_{m,n})} \, (X_{k+1,l}, X_{m+1,n})_{L(X)}$$

$$= \sum_{k,l} \sum_{m,n} (a_{k,l} - b_{k,l}) \overline{(a_{m,n} - b_{m,n})} \, (X_{k,l}, X_{m,n})_{L(X)}$$

$$= \left\| \sum_{k,l} a_{k,l} X_{k,l} - \sum_{k,l} b_{k,l} X_{k,l} \right\|_{L(X)}^2 = 0$$

Note that the third equality follows from stationarity. Therefore, U_1 is well defined on the collection of all finite linear combinations of $X_{m,n}$, $(m,n) \in \mathbb{Z}^2$. We also see that U_1 preserves the inner product.

$$\left(U_1 \left(\sum_{k,l} a_{k,l} X_{k,l} \right), U_1 \left(\sum_{k,l} b_{k,l} X_{k,l} \right) \right)_{L(X)}$$

$$= \left(\sum_{k,l} a_{k,l} X_{k+1,l}, \sum_{k,l} b_{k,l} X_{k+1,l} \right)_{L(X)}$$

$$= \sum_{k,l} \sum_{m,n} a_{k,l} \overline{b}_{k,l} \, (X_{k+1,l}, X_{m+1,n})_{L(X)} = \sum_{k,l} \sum_{m,n} a_{k,l} \overline{b}_{k,l} \, (X_{k,l}, X_{m,n})_{L(X)}$$

$$= \left(\sum_{k,l} a_{k,l} X_{k,l}, \sum_{k,l} b_{k,l} X_{k,l} \right)_{L(X)}.$$

Again, like above, the third equality follows from stationarity. Therefore, we have a linear operator U_1 that preserves the inner product, which is defined on a dense linear manifold of $L(X)$. Therefore, we can extend U_1 to all of $L(X)$. This extension, which we will also call U_1, is our desired unitary operator. It is left to show that U_1 is onto. To see this, let $\zeta \in L(X)$. Then, ζ may be written as the following limit.

$$\zeta = \lim_n \sum_{k,l} a_{k,l}^{(n)} X_{k,l}.$$

Using the same type of calculations as above, one can show that $\left(\sum_{k,l} a_{k,l}^{(n)} X_{k-1,l} \right)_n$ is a Cauchy sequence in $L(X)$ and hence has a limit. Let us call this limit η. It is straightforward to see that $U_1 \eta = \zeta$. Therefore, U_1 is onto.

Since U_1 and U_2 commute, it follows that they generate a commutative B^*-algebra. As such, the spectral-type representation theorem, given in the appendix, says that there exists a resolution of the identity $\mathcal{E}$ on $\mathcal{B}(\mathbb{T}^2)$, the Borel subsets of $\mathbb{T}^2$, the unit torus, such that

$$U_1^m U_2^n = \int_{[-\pi,\pi)} \int_{[-\pi,\pi)} e^{im\lambda+in\theta} \, d\mathcal{E}(e^{i\lambda}, e^{i\theta}), \quad ((m,n) \in \mathbb{Z}^2).$$

Defining $\mathcal{Z}_X : \mathcal{B}(\mathbb{T}^2) \to L(X)$, by

$$\mathcal{Z}_X(\Delta) = \mathcal{E}(\Delta)X_{0,0}, \quad (\Delta \in \mathcal{B}(\mathbb{T}^2)),$$

gives us an orthogonally scattered set function, see the appendix for details. It then follows that

$$X_{m,n} = U_1^m U_2^n X_{0,0} = \int_{[-\pi,\pi)} \int_{[-\pi,\pi)} e^{im\lambda+in\theta} \, d\mathcal{Z}_X(e^{i\lambda}, e^{i\theta}), \quad ((m,n) \in \mathbb{Z}^2),$$

which is called the spectral representation of $X_{m,n}$. We also observe that

$$r_X(m,n) = (U_1^m U_2^n X_{0,0}, X_{0,0}) = \int_{[-\pi,\pi)} \int_{[-\pi,\pi)} e^{im\lambda+in\theta} \, d\mathcal{E}_{X_0,X_0}(e^{i\lambda}, e^{i\theta}),$$

where $\mathcal{E}_{X_0,X_0}(\Delta) = (\mathcal{E}(\Delta)X_0, X_0)$, which we recall is a finite positive measure. Once again, the details are given in the appendix. This representation is called the spectral representation of the covariance function. The finite positive measure $\mathcal{E}_{X_0,X_0}$ is called the spectral measure of the random field $X_{m,n}, (m,n) \in \mathbb{Z}^2$. Henceforth, we will denote the spectral measure of $X_{m,n}, (m,n) \in \mathbb{Z}^2$ by F_X.

Let $L^2(\mathbb{T}^2, F_X)$ denote the collection of all measurable complex-valued functions defined on $\mathbb{T}^2$ and square integrable with respect to F_X. As always, we will identify functions that are equal $[F_X]$-a.e. We now show how one can transform a prediction problem from $L(X)$ to the function space $L^2(\mathbb{T}^2, F_X)$ and then back.

  Since $L(X)$ and $L^2(\mathbb{T}^2, F_X)$ are separable Hilbert spaces, there are many isomorphisms between them. We are interested in one particular isomorphism. Define $\mathcal{J} : L^2(\mathbb{T}^2, F_X) \to L(X)$ by

$$\mathcal{J}(e^{im\lambda + in\theta}) = X_{m,n}.$$

Extending this by linearity, we have for a finite linear combination that

$$\mathcal{J}\left(\sum_{k,l} a_{k,l} e^{ik\lambda + il\theta}\right) = \sum_{k,l} a_{k,l} X_{k,l}.$$

We now have that $\mathcal{J}$ is a linear mapping between the collection of all finite combinations of $\{e^{im\lambda + in\theta} : (m,n) \in \mathbb{Z}^2\}$ and the collection of all finite combinations of $\{X_{m,n} : (m,n) \in \mathbb{Z}^2\}$. This mapping is clearly onto. We also see that $\mathcal{J}$ preserves the inner product since,

$$\left(\mathcal{J}\left(\sum_{k,l} a_{k,l} e^{ik\lambda + il\theta}\right), \mathcal{J}\left(\sum_{m,n} b_{m,n} e^{im\lambda + in\theta}\right)\right)_{L(X)}$$

$$= \left(\sum_{k,l} a_{k,l} X_{k,l}, \sum_{m,n} b_{m,n} X_{m,n}\right)_{L(X)} = \sum_{k,l}\sum_{m,n} a_{k,l} \bar{b}_{m,n} (X_{k,l}, X_{m,n})_{L(X)}$$

$$= \sum_{k,l}\sum_{m,n} a_{k,l} \bar{b}_{m,n} r_X(k - m, l - n)$$

$$= \sum_{k,l}\sum_{m,n} a_{k,l} \bar{b}_{m,n} \int_{[-\pi,\pi)} \int_{[-\pi,\pi)} e^{i(k-m)\lambda + i(l-n)\theta} dF_X(e^{i\lambda}, e^{i\theta})$$

$$= \int_{[-\pi,\pi)} \int_{[-\pi,\pi)} \sum_{k,l} a_{k,l} e^{ik\lambda + il\theta} \overline{\sum_{m,n} b_{m,n} e^{im\lambda + in\theta}} dF_X(e^{i\lambda}, e^{i\theta})$$

$$= \left(\sum_{k,l} a_{k,l} e^{ik\lambda + il\theta}, \sum_{m,n} b_{m,n} e^{im\lambda + in\theta}\right)_{L^2(\mathbb{T}^2, F_X)}$$

and is therefore a one-to-one mapping. We now have that $\mathcal{J}$ is an isomorphism between two dense linear manifolds of $L^2(\mathbb{T}^2, F_X)$ and $L(X)$, respectively, and as such may be extended to an isomorphism, which we will still call $\mathcal{J}$, from $L^2(\mathbb{T}^2, F_X)$ to $L(X)$. We call this isomorphism our canonical isomorphism.

2.2   Examples

We now turn to some examples. In the following examples, we will use L_0^2 to denote the set of all $\zeta \in L^2(\Omega, \mathcal{F}, P)$ with the property that $E\zeta = 0$.

2.2.1 Random Field with Discrete Spectral Measure

Let $Z_1, ..., Z_N$ be an orthogonal collection of random variables in L_0^2. That is, $\text{cov}(Z_l, Z_k) = 0$ for all $k \neq l$. Now, let $\alpha_1, ..., \alpha_N, \beta_1, ..., \beta_N \in [-\pi, \pi)$ and define $X_{m,n} = \sum_{k=1}^{N} Z_k e^{im\alpha_k + in\beta_k}$, $(m, n) \in \mathbb{Z}^2$. X_n is a weakly stationary random field with covariance function $r_X(m, n) = \sum_{k=1}^{N} e^{im\alpha_k + in\beta_k} E|Z_k|^2$, $(m, n) \in \mathbb{Z}^2$. $X_{m,n}$ has spectral measure F_X that is a discrete measure concentrated at the points $(e^{i\alpha_1}, e^{i\beta_1}), ..., (e^{i\alpha_N}, e^{i\beta_N})$ with masses $F_X(\{(e^{i\alpha_k}, e^{i\beta_k})\}) = E|Z_k|^2$.

2.2.2 Product Random Field

Let Y_n, $n \in \mathbb{Z}$ and Z_n, $n \in \mathbb{Z}$ be two weakly stationary sequences. Further, suppose that $\sigma(Y_n : n \in \mathbb{Z})$ and $\sigma(Z_n : n \in \mathbb{Z})$ are independent. Define $X_{m,n} = Y_m Z_n$ for $(m, n) \in \mathbb{Z}^2$. Then, $X_{m,n}$ for $(m, n) \in \mathbb{Z}^2$ is a weakly stationary random field. Indeed, $EX_{m,n} = EY_m Z_n = EY_m EZ_n = 0$ for all $(m, n) \in \mathbb{Z}^2$ and $\text{cov}(X_{m,n}, X_{m',n'}) = EX_{m,n} \overline{X_{m',n'}} = EY_m Z_n \overline{Y_{m'} Z_{n'}} = EY_m \overline{Y_{m'}} Z_n \overline{Z_{n'}} = EY_m \overline{Y_{m'}} EZ_n \overline{Z_{n'}} = r_Y(m - m') r_Z(n - n')$. We see from this that $X_{m,n}$ for $(m, n) \in \mathbb{Z}^2$ is indeed a weakly stationary random field with covariance function $r_X(m, n) = r_Y(m) r_Z(n)$. We see from this that its spectral measure F is given by $dF(\lambda, \theta) = dF_Y(\lambda) dF_Z(\theta)$.

2.2.3 White Noise Random Field

In this example and henceforth, we will use σ^2 to denote Lebesgue measure on $\mathbb{T}^2$, normalized so that $\sigma^2(\mathbb{T}^2) = 1$.

Take $X_{m,n}$, $(m, n) \in \mathbb{Z}^2$ to be an orthonormal sequence of random variables in L_0^2. Then,

$$\text{cov}(X_{m,n}, X_{k,l}) = \delta_{(m,n)(k,l)} = \begin{cases} 1, & (m, n) = (k, l) \\ 0, & (m, n) \neq (k, l) \end{cases}.$$

This random field is often called white noise. It is weakly stationary with covariance function $r_X(m, n) = \delta_{(m,n)(0,0)}$, $(m, n) \in \mathbb{Z}^2$. The spectral measure for this random field is $F_X = \sigma^2$.

2.2.4 Moving Average Random Field

For $(m, n) \in \mathbb{Z}^2$, let

$$X_{m,n} = \sum_{(j,k) \in \mathbb{Z}^2} a_{j,k} \xi_{m-j, n-k},$$

where $(a_{m,n})_{(m,n)\in\mathbb{Z}^2}$ is a sequence of complex numbers with the property that $\sum\limits_{(m,n)\in\mathbb{Z}^2} |a_{m,n}|^2 < \infty$ and $\xi_{m,n}$, $(m,n) \in \mathbb{Z}^2$ is a white noise random field in $L^2(\Omega, \mathcal{F}, P)$. A straightforward calculation shows that $X_{m,n}$, $(m,n) \in \mathbb{Z}^2$ is a weakly stationary random field in $L^2(\Omega, \mathcal{F}, P)$.

We begin by examining the spectral measure for this random field. First recall that white noise is a stationary random field and as such has a spectral representation.

$$\xi_{m,n} = \int_{[-\pi,\pi)} \int_{[-\pi,\pi)} e^{im\lambda + in\theta} dZ_\xi(e^{i\lambda}, e^{i\theta}),$$

where Z_ξ is the orthogonally scattered set function associated with the random field $\xi_{m,n}$, $(m,n) \in \mathbb{Z}^2$. We pointed out in the previous example that F_ξ, the spectral measure for $\xi_{m,n}$, $(m,n) \in \mathbb{Z}^2$, is normalized Lebesgue measure on $\mathbb{T}^2$. That is, $dF_\xi(e^{i\lambda}, e^{i\theta}) = d\sigma^2(e^{i\lambda}, e^{i\theta})$. We now calculate the covariance function for $X_{m,n}$, $(m,n) \in \mathbb{Z}^2$.

$$r_X(m,n) = (X_{m,n}, X_{0,0})_{L^2(\Omega,\mathcal{F},P)}$$

$$= \left(\sum_{(j,k)\in\mathbb{Z}^2} a_{j,k}\, \xi_{m-j,n-k},\ \sum_{(j,k)\in\mathbb{Z}^2} a_{j,k}\, \xi_{-j,-k} \right)_{L^2(\Omega,\mathcal{F},P)}$$

$$= \int_{[-\pi,\pi)} \int_{[-\pi,\pi)} \sum_{(j,k)\in\mathbb{Z}^2} a_{j,k} e^{i(m-j)\lambda + i(n-k)\theta} \overline{\sum_{(j,k)\in\mathbb{Z}^2} a_{j,k} e^{-ij\lambda - ik\theta}} \, dF_\xi(e^{i\lambda}, e^{i\theta})$$

$$= \int_{[-\pi,\pi)} \int_{[-\pi,\pi)} \left| \sum_{(j,k)\in\mathbb{Z}^2} a_{j,k} e^{-ij\lambda - ik\theta} \right|^2 e^{im\lambda + in\theta} \, d\sigma^2(e^{i\lambda}, e^{i\theta}).$$

Therefore, we see that the spectral measure of $X_{m,n}$, $(m,n) \in \mathbb{Z}^2$, is absolutely continuous with respect to σ^2, normalized Lebesgue measure on $\mathbb{T}^2$, with density $f_X(e^{i\lambda}, e^{i\theta}) = \left| \sum\limits_{(j,k)\in\mathbb{Z}^2} a_{j,k}\, e^{-ij\lambda - ik\theta} \right|^2 = \left| \sum\limits_{(j,k)\in\mathbb{Z}^2} \overline{a_{j,k}}\, e^{ij\lambda + ik\theta} \right|^2$. We now record this observation as a theorem.

Theorem 2.2.1. *Suppose that* $X_{m,n}$, $(m,n) \in \mathbb{Z}^2$ *is a moving average random field; that is, there exists a sequence of complex numbers* $(a_{m,n})_{(m,n)\in\mathbb{Z}^2}$ *with the property that* $\sum\limits_{(m,n)\in\mathbb{Z}^2} |a_{m,n}|^2 < \infty$ *and a white noise random field* $\xi_{m,n}$, $(m,n) \in \mathbb{Z}^2$ *in* $L^2(\Omega, \mathcal{F}, P)$ *such that*

$$X_{m,n} = \sum_{(j,k)\in\mathbb{Z}^2} a_{j,k}\, \xi_{m-j,n-k},$$

in $L^2(\Omega, \mathcal{F}, P)$, *for all* $(m,n) \in \mathbb{Z}^2$. *Then,* $X_{m,n}$, $(m,n) \in \mathbb{Z}^2$ *is a weakly*

stationary random field that has a spectral measure that is absolutely contin-
uous with respect to σ^2, *normalized Lebesgue measure on* $\mathbb{T}^2$, *with density*
$f_X(e^{i\lambda}, e^{i\theta}) = |\varphi(e^{i\lambda}, e^{i\theta})|^2$, *where* $\varphi \in L^2(\mathbb{T}^2, \sigma^2)$ *and* $\hat{\varphi}(j, k) = \overline{a_{j,k}}$ *for all*
$(j, k) \in \mathbb{Z}^2$, *where* $\hat{\varphi}(j, k)$ *denotes the* (j, k) *Fourier coefficient of* φ.

Now, let us examine the situation from the other direction. Suppose that
$X_{m,n}$, $(m, n) \in \mathbb{Z}^2$ is a weakly stationary random field with spectral measure
$dF_X(e^{i\lambda}, e^{i\theta}) = |\varphi(e^{i\lambda}, e^{i\theta})|^2 \, d\sigma^2(e^{i\lambda}, e^{i\theta})$, where $\varphi \in L^2(\mathbb{T}^2, \sigma^2)$. As such, φ
is equal to its Fourier series in $L^2(\mathbb{T}^2, \sigma^2)$. That is,

$$\varphi(e^{i\lambda}, e^{i\theta}) = \sum_{(j,k)\in\mathbb{Z}^2} \hat{\varphi}(j, k) e^{ij\lambda+ik\theta} \text{ in } L^2(\mathbb{T}^2, \sigma^2).$$

We now define the linear transformation $W : L(X) \to L^2(\mathbb{T}^2, \sigma^2)$ by
$W(X_{m,n}) = e^{im\lambda+in\theta}\overline{\varphi}(e^{i\lambda}, e^{i\theta})$. W is an isometry and as such, we may
identify $L(X)$ and $W(L(X))$, as well as, $X_{m,n}$ and $e^{im\lambda+in\theta}\overline{\varphi}(e^{i\lambda}, e^{i\theta})$.
$e^{im\lambda+in\theta}\overline{\varphi}(e^{i\lambda}, e^{i\theta})$ has a moving average representation since

$$e^{im\lambda+in\theta}\overline{\varphi}(e^{i\lambda}, e^{i\theta}) = \sum_{(j,k)\in\mathbb{Z}^2} \overline{\hat{\varphi}}(j, k) e^{i(m-j)\lambda+i(n-k)\theta} \text{ in } L^2(\mathbb{T}^2, \sigma^2).$$

So, we may conclude the $X_{m,n}$, $(m, n) \in \mathbb{Z}^2$ has a moving average representation of the form

$$X_{m,n} = \sum_{(j,k)\in\mathbb{Z}^2} \overline{\hat{\varphi}(j, k)}\xi_{m-j,n-k},$$

for some white noise random field $(\xi_{m,n})_{(m,n)\in\mathbb{Z}^2}$ contained in some Hilbert
space that contains $L(X)$ as a subspace. A natural question at this point might
be: when is this white noise sequence contained in $L(X)$? Let us examine that
question. Let

$$L(\xi) = \overline{\text{span}}\left\{\xi_{m,n} : (m, n) \in \mathbb{Z}^2\right\}.$$

Now, define the linear transformation $W^{\#} : L(\xi) \to L^2(\mathbb{T}^2, \sigma^2)$ by
$W^{\#}(\xi_{m,n}) = e^{im\lambda+in\theta}$. $W^{\#}$ is clearly an isomorphism. It follows from the
moving average representation of $X_{m,n}$, $(m, n) \in \mathbb{Z}^2$, that $L(X) \subseteq L(\xi)$.
Again from this moving average representation, we see that

$$W^{\#}(X_{m,n}) = e^{im\lambda+in\theta}\overline{\varphi}(e^{i\lambda}, e^{i\theta}).$$

It follows then that

$$W^{\#}(L(X)) = \overline{\text{span}}\left\{e^{im\lambda+in\theta}\overline{\varphi}(e^{i\lambda}, e^{i\theta}) : (m, n) \in \mathbb{Z}^2\right\},$$

which is a doubly invariant subspace of $L^2(\mathbb{T}^2, \sigma^2)$. By Theorem 3.6.1,
all doubly invariant subspaces of $L^2(\mathbb{T}^2, \sigma^2)$ are of the form $1_E L^2(\mathbb{T}^2, \sigma^2)$,
where E is a Lebesgue measurable subset of $\mathbb{T}^2$. In our particular case,
$E = \{(e^{i\lambda}, e^{i\theta}) \in \mathbb{T}^2 : \varphi(e^{i\lambda}, e^{i\theta}) \neq 0\}$. From this observation, we see that if
E^c, the complement of E, has Lebesgue measure zero, then $W^{\#}(L(X)) =$

$L^2(\mathbb{T}^2, \sigma^2)$. This together with the fact that $W^\#$ is an isomorphism gives us that $L(X) = L(\xi)$ and therefore, $\xi_{m,n} \in L(X)$ for all $(m,n) \in \mathbb{Z}^2$. If, on the other hand, the Lebesgue measure of E^c is positive, then $L(X) \subsetneq L(\xi)$. So, there exists a $(j,k) \in \mathbb{Z}^2$ such that $\xi_{j,k} \notin L(X)$. An important observation to make here, if not already observed, is that $|\varphi(e^{i\lambda}, e^{i\theta})|^2$ is the density of the spectral measure of our random field $X_{m,n}$, $(m,n) \in \mathbb{Z}^2$. Therefore, whether or not our spectral density vanishes on a set of positive measure impacts the location of our white noise random field.

We will now record these observations as a theorem.

Theorem 2.2.2. *Suppose that $X_{m,n}$, $(m,n) \in \mathbb{Z}^2$ is a weakly stationary random field with spectral measure $dF_X(e^{i\lambda}, e^{i\theta}) = |\varphi(e^{i\lambda}, e^{i\theta})|^2 \, d\sigma^2(e^{i\lambda}, e^{i\theta})$, where $\varphi \in L^2(\mathbb{T}^2, \sigma^2)$. Then, $X_{m,n}$, $(m,n) \in \mathbb{Z}^2$ may be represented as a moving average random field; that is,*

$$X_{m,n} = \sum_{(j,k) \in \mathbb{Z}^2} \overline{\hat{\varphi}(j,k)} \xi_{m-j,n-k},$$

in some Hilbert space $\mathcal{H}$ containing the white noise random field $\xi_{m,n}$, $(m,n) \in \mathbb{Z}^2$ and having $L(X)$ as a subspace. If $\varphi \neq 0$ $[\sigma^2]$-a.e., then our white noise random field is contained in $L(X)$.

2.3   Regularity and Singularity

Let $X_{m,n}$, $(m,n) \in \mathbb{Z}^2$ be a weakly stationary random field. Let $L(X)$ be as defined previously. Define

$$L^1(X : m) = \overline{\text{span}}\{X_{j,k} : j \leq m, k \in \mathbb{Z}\},$$

$$L^2(X : n) = \overline{\text{span}}\{X_{j,k} : j \in \mathbb{Z}, k \leq n\},$$

$$L^1(X : -\infty) = \bigcap_m L^1(X : m)$$

and

$$L^2(X : -\infty) = \bigcap_n L^2(X : n).$$

Note that $L^j(X : -\infty) \subseteq L^j(X : n) \subseteq L^j(X : n+1) \subset L(X)$, for $j = 1, 2$.

We now define some concepts of regularity and singularity in the context of random fields. A second order random field $X_{m,n}$, $(m,n) \in \mathbb{Z}^2$ is called

- horizontally regular if $L^1(X : -\infty) = \{0\}$,

- vertically regular if $L^2(X : -\infty) = \{0\}$,

- horizontally singular if $L(X) = L^1(X : -\infty)$, and

- vertically singular if $L(X) = L^2(X : -\infty)$.

We leave this section with a few observations. First, since

$$U_j L^j(X : m) = L^j(X : m+1), \quad (j = 1, 2),$$

it follows that if $L^j(X : m) = L^j(X : m + 1)$, $(j = 1, 2)$ for some $m \in \mathbb{Z}$ then $L^j(X : m) = L^j(X : m+1)$, $(j = 1, 2)$ for all $m \in \mathbb{Z}$. Finally, if $X_{m,n} \in L^1(X : m - 1)$, for some $m \in \mathbb{Z}$ then $L^1(X : m) = L^1(X : m + 1)$. Similarly, if $X_{m,n} \in L^2(X : n - 1)$, for some $n \in \mathbb{Z}$ then $L^2(X : n) = L^2(X : n + 1)$. Combining these observations, we see that if $X_{m,n} \in L^1(X : m - 1)$, for some $m \in \mathbb{Z}$ then $X_{m,n}$, $(m, n) \in \mathbb{Z}^2$ is horizontally singular. Similarly, if $X_{m,n} \in L^2(X : n - 1)$, for some $n \in \mathbb{Z}$ then $X_{m,n}$, $(m, n) \in \mathbb{Z}^2$ is vertically singular.

2.4 Examples

To get a better feel for the concepts of regularity and singularity, we look at a few examples. Recall that L_0^2 denotes the set of all $\zeta \in L^2(\Omega, \mathcal{F}, P)$ with the property that $E\zeta = 0$.

2.4.1 Horizontally and Vertically Singular

Let $\zeta \in L_0^2$, $\alpha, \beta \in [-\pi, \pi)$ and define $X_{m,n} = \zeta e^{im\alpha + in\beta}$, $(m, n) \in \mathbb{Z}^2$. It is straightforward to check that $X_{m,n}$ is a weakly stationary random field with covariance function $r_X(m, n) = e^{im\alpha + in\beta} E|\zeta|^2$, $(m, n) \in \mathbb{Z}^2$. $X_{m,n}$ has spectral measure F_X that is a discrete measure concentrated on the point $(e^{i\alpha}, e^{i\beta})$ with mass $F_X(\{(e^{i\alpha}, e^{i\beta})\}) = E|\zeta|^2$.

To see that $X_{m,n}$, $(m, n) \in \mathbb{Z}^2$ is horizontally singular, note that $X_{m,n} = e^{i\alpha} X_{m-1,n}$. Therefore, $X_{m,n} \in L^1(X : m - 1)$ and by our observations from the last section, we conclude that $X_{m,n}$, $(m, n) \in \mathbb{Z}^2$ is horizontally singular. Similarly, observing that $X_{m,n} = e^{i\beta} X_{m,n-1}$ gives us that $X_{m,n} \in L^2(X : n - 1)$ and so we may conclude that $X_{m,n}$, $(m, n) \in \mathbb{Z}^2$ is vertically singular.

2.4.2 Horizontally Regular and Vertically Singular

For $(m, n) \in \mathbb{Z}^2$, define $X_{m,n} = \xi_m$, where $(\xi_m)_m$ is an orthonormal sequence in L_0^2. Then, $X_{m,n}$ is a weakly stationary random field with covariance function $r_X(m, n) = \delta_{m0}$. $X_{m,n}$ has spectral measure $F_X = \sigma \otimes \mu$, where σ is Lebesgue measure on $\mathbb{T}$ normalized so that $\sigma(\mathbb{T}) = 1$ and μ is a discrete measure concentrated on the point $1 \in \mathbb{T}$ with mass $\mu(\{1\}) = 1$. We start by showing that $X_{m,n} = \xi_m$, $(m, n) \in \mathbb{Z}^2$ is horizontally regular.

First observe that

$$L^1(X:m) = \overline{span}\{X_{k,l} : k \leq m, l \in \mathbb{Z}\} = \overline{span}\{\xi_k : k \leq m\}$$

and

$$L(X) = \overline{span}\{X_{k,l} : (k,l) \in \mathbb{Z}^2\} = \overline{span}\{\xi_k : k \in \mathbb{Z}\}.$$

Now, since $(\xi_m)_m$ is an orthonormal sequence in L_0^2, it follows that $(\xi_m)_m$ is an orthonormal basis for $L(X)$ and $(\xi_k)_{k \leq m}$ is an orthonormal basis for $L^1(X:m)$. If $Y \in L^1(X:-\infty)$, then $Y \in L^1(X:m)$ for all $m \in \mathbb{Z}$ and $Y \in L(X)$. Therefore, Y has the following unique representations. Because $Y \in L(X)$, it can be written uniquely as

$$Y = \sum_k c_k \xi_k$$

and because $Y \in L^1(X:m)$ for all $m \in \mathbb{Z}$ it can be written uniquely as

$$Y = \sum_{k \leq m} a_{m,k} \xi_k.$$

By the uniqueness of the representations, it follows that $c_k = 0$ for all $k \in \mathbb{Z}$ and hence, $Y = 0$. Therefore, $L^1(X:-\infty) = \{0\}$. Hence, $X_{m,n} = \xi_m$, $(m,n) \in \mathbb{Z}^2$ is horizontally regular.

We now show that $X_{m,n} = \xi_m$, $(m,n) \in \mathbb{Z}^2$ is vertically singular. We can see this quickly by observing that

$$L^2(X:n) = \overline{span}\{X_{k,l} : k \in \mathbb{Z}, l \leq n\} = \overline{span}\{\xi_k : k \in \mathbb{Z}\} = L(X).$$

It follows that $L^2(X:-\infty) = L(X)$ and hence, by definition, $X_{m,n} = \xi_m$, $(m,n) \in \mathbb{Z}^2$ is vertically singular.

2.4.3 Horizontally and Vertically Regular

Let $\xi_{m.n}$, $(m,n) \in \mathbb{Z}^2$ be a white noise random field. We will show that $\xi_{m.n}$, $(m,n) \in \mathbb{Z}^2$ is horizontally regular. An analogous argument will show that $\xi_{m.n}$, $(m,n) \in \mathbb{Z}^2$ is vertically regular.

First recall that

$$L^1(X:m) = \overline{span}\{\xi_{k,l} : k \leq m, l \in \mathbb{Z}\}$$

and

$$L(X) = \overline{span}\{\xi_{k,l} : (k,l) \in \mathbb{Z}^2\}.$$

Now, since $(\xi_m)_m$ is an orthonormal sequence in L_0^2, it follows that $(\xi_{m,n})_{(m,n)}$ is an orthonormal basis for $L(X)$ and $(\xi_{k,l})_{l \in \mathbb{Z}, k \leq m}$ is an orthonormal basis for $L^1(X:m)$. If $Y \in L^1(X:-\infty)$, then $Y \in L^1(X:m)$ for all $m \in \mathbb{Z}$ and

$Y \in L(X)$. Therefore, Y has the following unique representations. Because $Y \in L(X)$, it can be written uniquely as

$$Y = \sum_{k,l} c_{k,l} \xi_{k,l}$$

and because $Y \in L^1(X : m)$ for all $m \in \mathbb{Z}$ it can be written uniquely as

$$Y = \sum_{l \in \mathbb{Z}, k \leq m} a_{k,l}^{(m)} \xi_{k,l}.$$

By the uniqueness of the representations, it follows that $c_{k,l} = 0$ for all $(k,l) \in \mathbb{Z}^2$ and hence, $Y = 0$. Therefore, $L^1(X : -\infty) = \{0\}$. Hence, $X_{m,n} = \xi_m$, $(m,n) \in \mathbb{Z}^2$ is horizontally regular.

2.5 Horizontal and Vertical Wold Decomposition

We start by proving a few preliminary lemmas. Let

$$W_n^j = L^j(X : n) \ominus L^j(X : n-1), \text{ for } j = 1, 2.$$

Lemma 2.5.1. $L^j(X : n) = \left(\sum_{k=0}^{\infty} \oplus W_{n-k}^j \right) \bigoplus L^j(X : -\infty), \text{ for } j = 1, 2.$

Proof. Henceforth, P_M will denote the projection on $L(X)$ onto the (closed) subspace M. See the appendix for some background material on projections used in this proof. For $j = 1, 2$,

$$
\begin{aligned}
P_{L^j(X:n)} &= \left(P_{L^j(X:n)} - P_{L^j(X:n-1)} \right) + P_{L^j(X:n-1)} \\
&= P_{W_n^j} + P_{L^j(X:n-1)} \\
&= P_{W_n^j} + \left(P_{L^j(X:n-1)} - P_{L^j(X:n-2)} \right) + P_{L^j(X:n-2)} \\
&= P_{W_n^j} + P_{W_{n-1}^j} + P_{L^j(X:n-2)} \\
&= \cdots \\
&= \sum_{k=0}^{l} P_{W_{n-k}^j} + P_{L^j(X:n-l-1)}
\end{aligned}
$$

Letting l go to infinity, we get

$$P_{L^j(X:n)} = \sum_{k=0}^{\infty} P_{W_{n-k}^j} + P_{L^j(X:-\infty)}. \tag{2.1}$$

It then follows that

$$L^j(X:n) = \left(\sum_{k=0}^{\infty} \oplus W_{n-k}^j \right) \bigoplus L^j(X:-\infty)$$

as desired.   □

Lemma 2.5.2. $L(X) = \left(\displaystyle\sum_{k=-\infty}^{\infty} \oplus W_k^j \right) \bigoplus L^j(X:-\infty)$, *for* $j = 1, 2$.

Proof. We may rewrite, (2.1) as

$$P_{L^j(X:n)} = \sum_{k=-\infty}^{n} P_{W_k^j} + P_{L^j(X:-\infty)}.$$

Now, letting n go to infinity, we get

$$P_{L(X)} = \sum_{k=-\infty}^{\infty} P_{W_k^j} + P_{L^j(X:-\infty)}.$$

It then follows that

$$L(X) = \left(\sum_{k=-\infty}^{\infty} \oplus W_k^j \right) \bigoplus L^j(X:-\infty)$$

as desired.   □

Lemma 2.5.3. *For* $j = 1, 2$, *we have that* $L^j(X:-\infty) = \{0\}$ *if and only if* $\displaystyle\lim_{l \to -\infty} P_{L^j(X:l)} X_{m,n} = 0$ *for all* $(m, n) \in \mathbb{Z}^2$.

Proof. This follows immediately from the fact that $P_{L^j(X:l)} \downarrow P_{L^j(X:-\infty)}$ strongly.   □

The following result is one-sided and will be presented in the horizontal direction. An analogous result holds in the vertical direction. Straightforward adjustments to statements and proofs should easily give the analogous result in the vertical direction.

Before we state and prove the Horizontal Wold Decomposition, we prove two lemmas. Recall that U_1 is the unitary operator on $L(X)$ with the property that $U_1 X_{m,n} = X_{m+1,n}$.

Lemma 2.5.4. $U_1 P_{L^1(X:n)} = P_{L^1(X:n+1)} U_1$ *on* $L(X)$.

Proof. Let $Y \in L(X)$. Then, Y has a unique decomposition

$$Y = P_{L^1(X:n)} Y + \left(Y - P_{L^1(X:n)} Y \right).$$

Note that $P_{L^1(X:n)}Y \in L^1(X:n)$ and $\left(Y - P_{L^1(X:n)}Y\right) \in L(X) \ominus L^1(X:n)$. Now, applying U_1 to both sides, we get that

$$U_1 Y = U_1 P_{L^1(X:n)}Y + \left(U_1 Y - U_1 P_{L^1(X:n)}Y\right).$$

This is a unique decomposition of $U_1 Y$ with $U_1 P_{L^1(X:n)}Y \in L^1(X:n+1)$ and $\left(U_1 Y - U_1 P_{L^1(X:n)}Y\right) \in L(X) \ominus L^1(X:n+1)$. Note that using the same approach as that with Y, we get the unique decomposition

$$U_1 Y = P_{L^1(X:n+1)}U_1 Y + \left(U_1 Y - P_{L^1(X:n+1)}U_1 Y\right),$$

where $P_{L^1(X:n+1)}U_1 Y \in L^1(X:n+1)$ and $\left(U_1 Y - P_{L^1(X:n+1)}U_1 Y\right) \in L(X) \ominus L^1(X:n+1)$. The uniqueness of these decompositions shows that

$$U_1 P_{L^1(X:n)}Y = P_{L^1(X:n+1)}U_1 Y.$$

Therefore, $U_1 P_{L^1(X:n)} = P_{L^1(X:n+1)}U_1$ on $L(X)$, as desired. $\qquad \square$

Lemma 2.5.5. $U_1 P_{L^1(X:-\infty)} = P_{L^1(X:-\infty)}U_1$ on $L(X)$.

Proof. Recall that $P_{L^1(X:n)}$ converges to $P_{L^1(X:-\infty)}$ strongly. Therefore, $P_{L^1(X:n)}U_1$ converges to $P_{L^1(X:-\infty)}U_1$ strongly. Since U_1 is unitary, it also follows that $U_1 P_{L^1(X:n)}$ converges to $U_1 P_{L^1(X:-\infty)}$ strongly. Hence, for $Y \in L(X)$, we have

$$\|U_1 P_{L^1(X:-\infty)}Y - P_{L^1(X:-\infty)}U_1 Y\|_{L(X)}$$

$$= \|U_1 P_{L^1(X:-\infty)}Y - U_1 P_{L^1(X:n)}Y + P_{L^1(X:n+1)}U_1 Y - P_{L^1(X:-\infty)}U_1 Y\|_{L(X)}$$

$$\leq \|U_1 P_{L^1(X:-\infty)}Y - U_1 P_{L^1(X:n)}Y\|_{L(X)} + \|P_{L^1(X:n+1)}U_1 Y - P_{L^1(X:-\infty)}U_1 Y\|_{L(X)}$$

$$\longrightarrow 0 \text{ as } n \longrightarrow \infty.$$

Therefore, $U_1 P_{L^1(X:-\infty)} = P_{L^1(X:-\infty)}U_1$ on $L(X)$. $\qquad \square$

Theorem 2.5.1 (Horizontal Wold Decomposition). *Suppose* $X_{m,n}$, $(m,n) \in \mathbb{Z}^2$ *is a weakly stationary sequence. Then,* $X_{m,n}$ *may be written as*

$$X_{m,n} = X_{m,n}^r + X_{m,n}^s, \qquad (m,n) \in \mathbb{Z}^2,$$

where

    1. $X_{m,n}^r$, $(m,n) \in \mathbb{Z}^2$ *is a horizontally regular weakly stationary random field.*

    2. $X_{m,n}^s$, $(m,n) \in \mathbb{Z}^2$ *is a horizontally singular weakly stationary random field.*

    3. $L^1(X^s:m) \subseteq L^1(X:m)$, $L^1(X^r:m) \subseteq L^1(X:m)$ *and* $L^1(X^r:m) \perp L^1(X^s:m)$ *for all* $m \in \mathbb{Z}$.

    4. $L^1(X:m) = L^1(X^r:m) \oplus L^1(X^s:m)$ *for all* $m \in \mathbb{Z}$.

    5. The decomposition satisfying these conditions is unique.

Proof. Let $X_{m,n}^s = P_{L^1(X:-\infty)}X_{m,n}$ and $X_{m,n}^r = X_{m,n} - X_{m,n}^s$ for all $(m,n) \in \mathbb{Z}^2$. It then follows from their definitions and Lemma 2.5.1 that:

    1. $X_{m,n} = X_{m,n}^r + X_{m,n}^s$ for all $(m,n) \in \mathbb{Z}^2$.

    2. $X_{m,n}^r \perp X_{k,l}^s$ for all $(m,n), (k,l) \in \mathbb{Z}^2$. Therefore, $L^1(X^r : m) \perp$ $L^1(X^s : m)$ for all $m \in \mathbb{Z}$.

    3. $L^1(X : m) \subseteq L^1(X^r : m) + L^1(X^s : m)$ for all $m \in \mathbb{Z}$.

Again, using Lemma 2.5.1, we also see that $L^1(X^s : m) \subseteq L^1(X : m)$ and $L^1(X^r : m) \subseteq L^1(X : m)$. It then follows that $L^1(X^r : m) + L^1(X^s : m) \subseteq L(X : m)$ for all $m \in \mathbb{Z}$. Putting this observation together with observation 3 above, we get that

$$L^1(X : m) = L^1(X^r : m) \oplus L^1(X^s : m) \text{ for all } m \in \mathbb{Z}.$$

    Using Lemma 2.5.5, we can show that X_n^s, $n \in \mathbb{Z}$ are stationary sequences. Indeed,

$$U_1 X_{m,n}^s = U_1 P_{L(X:-\infty)}X_{m,n} = P_{L(X:-\infty)}U_1 X_{m,n} = P_{L(X:-\infty)}X_{m+1,n} = X_{m+1,n}^s.$$

The stationarity of X_n^s, $n \in \mathbb{Z}$ can be used to show the stationarity of X_n^r, $n \in \mathbb{Z}$.

$$U_1 X_{m,n}^r = U_1(X_{m,n} - X_{m,n}^s) = U_1 X_{m,n} - U_1 X_{m,n}^s = X_{m+1,n} - X_{m+1,n}^s = X_{m+1,n}^r.$$

    We will now show that $X_{m,n}^r$, $(m,n) \in \mathbb{Z}^2$ is horizontally regular, and $X_{m,n}^s$, $(m,n) \in \mathbb{Z}^2$ is horizontally singular. To see that $X_{m,n}^s$, $(m,n) \in \mathbb{Z}^2$ is horizontally singular, note that $L^1(X^s : m) = L^1(X : m) \cap L^1(X : -\infty) = L^1(X : -\infty)$. This follows from the definition of $X_{m,n}^s$ and Lemma 2.5.1. It then follows that $L(X^s : -\infty) = L(X^s)$. That is, $X_{m,n}^s$, $(m,n) \in \mathbb{Z}^2$ is horizontally singular. To see that $X_{m,n}^r$, $(m,n) \in \mathbb{Z}^2$ is horizontally regular, note that by our previous observation that

$$L^1(X^r : m) = L^1(X : m) \ominus L^1(X^s : m) \text{ for all } m \in \mathbb{Z}.$$

Our previous observation then gives

$$L^1(X^r : m) = L^1(X : m) \ominus L^1(X : -\infty) \text{ for all } m \in \mathbb{Z}.$$

It then follows that $L^1(X^r : -\infty) = \{0\}$. That is, $X_{m,n}^r$, $(m,n) \in \mathbb{Z}^2$ is horizontally regular, as desired.

    It is left to show that this decomposition is unique. To see this, we employ property 3 of the Wold Decomposition, which states that if $X_{m,n}$ has another Wold Decomposition $X_{m,n} = \tilde{X}_{m,n}^r + \tilde{X}_{m,n}^s$ then $L^1(\tilde{X}^s : m) \subseteq L^1(X : m)$ and therefore, $L^1(\tilde{X}^s : -\infty) \subseteq L^1(X : -\infty)$. Since $\tilde{X}_{m,n}^s$ is horizontally singular, we may conclude that $L(\tilde{X}^s) \subseteq L^1(X : -\infty)$. Now using property 4 of the Wold Decomposition along with the fact that $\tilde{X}^r_{m,n}$ is horizontally regular, we see that $L(\tilde{X}^s) = L^1(X : -\infty)$. Therefore, the following calculation yields the uniqueness of this decomposition.

$$X^s_{m,n} = P_{L^1(X:-\infty)} X_{m,n} = P_{L(\tilde{X}^s)} X_{m,n} = \tilde{X}^s_{m,n}.$$

$\square$

2.6 Regularity and the Spectral Measure

We now find conditions on F_X, the spectral measure of $X_{m,n}$, $(m,n) \in \mathbb{Z}^2$, such that $X_{m,n}$, $(m,n) \in \mathbb{Z}^2$ is horizontally regular. Note that we will only investigate horizontally, but analogous results will hold vertically.

We now defined $\nu^1_{m,n}$, the horizontal innovations of $X_{m,n}$, by

$$\nu^1_{m,n} = X_{m,n} - P_{L^1(X:m-1)} X_{m,n}.$$

A straightforward calculation shows that $\nu^1_{m,n} = U^m_1 U^n_2 \nu^1_{0,0}$ for all $(m,n) \in \mathbb{Z}^2$. Therefore, $\nu^1_{m,n}$, $(m,n) \in \mathbb{Z}^2$ is a weakly stationary random field. Let $a(e^{i\lambda}, e^{i\theta}) = \mathcal{J}^{-1}(\nu^1_{0,0})$. Then,

$$\nu^1_{0,0} = \int_{[-\pi,\pi)} \int_{[-\pi,\pi)} a(e^{i\lambda}, e^{i\theta}) \, dZ_X(e^{i\lambda}, e^{i\theta})$$

and therefore,

$$\nu^1_{m,n} = \int_{[-\pi,\pi)} \int_{[-\pi,\pi)} e^{im\lambda + in\theta} a(e^{i\lambda}, e^{i\theta}) \, dZ_X(e^{i\lambda}, e^{i\theta}).$$

From this equation, we see that the spectral measure for $\nu^1_{m,n}$ is

$$F_{\nu^1}(\Delta) = \int \int_\Delta |a(e^{i\lambda}, e^{i\theta})|^2 \, dF_X(e^{i\lambda}, e^{i\theta}).$$

Therefore, F_{ν^1} is absolutely continuous with respect to F_X. Now, if $X_{m,n}$, $(m,n) \in \mathbb{Z}^2$ is horizontally regular, then by Lemma 2.5.2 we can conclude that $\nu^1_{m,n} \neq 0$ for all $(m,n) \in \mathbb{Z}^2$ and $L(\nu^1) = L(X)$. Let $\tilde{\mathcal{J}}$ be the canonical isometry from $L^2(F_{\nu^1})$ to $L(\nu^1)$ that takes $e^{im\lambda + in\theta}$ to $\nu^1_{m,n}$ and let $\tilde{a}(e^{i\lambda}, e^{i\theta}) = \tilde{\mathcal{J}}^{-1}(X_{0,0})$. Therefore,

$$X_{m,n} = \int_{[-\pi,\pi)} \int_{[-\pi,\pi)} e^{im\lambda + in\theta} \tilde{a}(e^{i\lambda}, e^{i\theta}) \, dZ_{\nu^1}(e^{i\lambda}, e^{i\theta}).$$

Hence,

$$F_X(\Delta) = \int \int_\Delta |\tilde{a}(e^{i\lambda}, e^{i\theta})|^2 \, dF_{\nu^1}(e^{i\lambda}, e^{i\theta}).$$

So we see that F_X is absolutely continuous with respect to F_{ν^1}. Therefore, we just proved the following lemma.

Lemma 2.6.1. *If the weakly stationary random field $X_{m,n}$, $(m,n) \in \mathbb{Z}^2$ is horizontally regular, then $F_X \equiv F_{\nu^1}$, where this symbolism means that F_X and F_{ν^1} are mutually absolutely continuous.*

Let G be any measure defined on $\mathcal{B}(\mathbb{T}^2)$. We define the marginal of G by

$$G_2(\Delta) = G(\mathbb{T} \times \Delta),$$

where Δ is in $\mathcal{B}(\mathbb{T})$.

Lemma 2.6.2. *If the weakly stationary random field $X_{m,n}$, $(m,n) \in \mathbb{Z}^2$ is horizontally regular, then the weakly stationary sequence $\nu^1_{0,n}$, $n \in \mathbb{Z}$ has spectral measure $F_{\nu^1 2}$.*

Note that this lemma remains true if we replace $\nu^1_{0,n}$, $n \in \mathbb{Z}$ with $\nu^1_{m,n}$, $n \in \mathbb{Z}$ for any fix $m \in \mathbb{Z}$. We now give the proof of this lemma.

Proof. $E\nu^1_{0,n}\overline{\nu^1_{0,0}} = \int_{[-\pi,\pi)} \int_{[-\pi,\pi)} e^{i0\lambda+in\theta} \, dF_{\nu^1}(e^{i\lambda}, e^{i\theta})$

$= \int_{[-\pi,\pi)} e^{in\theta} \, dF_{\nu^1 2}(e^{i\theta}).$ □

Lemma 2.6.3. *If the weakly stationary random field $X_{m,n}$, $(m,n) \in \mathbb{Z}^2$ is horizontally regular, then the spectral measure of $\nu^1_{m,n}$, $(m,n) \in \mathbb{Z}^2$ is $\sigma \otimes F_{\nu^1 2}$, where σ is Lebesgue measure normalized so that $\sigma(\mathbb{T}) = 1$.*

Proof. First note that by definition,

$$E\nu^1_{m,n}\overline{\nu^1_{0,0}} = \begin{cases} 0 & \text{if } m \neq 0, \\ \int_{[-\pi,\pi)} e^{in\theta} \, dF_{\nu^1 2}(e^{i\theta}) & \text{if } m = 0 \text{ (by the previous lemma).} \end{cases}$$

It is a straightforward calculation to see that the following double integral provides the same values.

$$\int_{[-\pi,\pi)} \int_{[-\pi,\pi)} e^{im\lambda+in\theta} \, d\sigma(e^{i\lambda}) dF_{\nu^1 2}(e^{i\theta})$$

The lemma then follows by the uniqueness of the spectral measure.   □

Lemma 2.6.4. *If the weakly stationary random field $X_{m,n}$, $(m,n) \in \mathbb{Z}^2$ is horizontally regular, then $F_X \equiv \sigma \otimes F_{\nu^1 2} \equiv \sigma \otimes F_{X2}$.*

Proof. We have already observed that $F_X \equiv F_{\nu^1} = \sigma \otimes F_{\nu^1 2}$. It follows from this that $F_{X2} \equiv F_{\nu^1 2}$ giving our desired result.   □

Theorem 2.6.1. *A weakly stationary random field* $X_{m,n}$, $(m,n) \in \mathbb{Z}^2$ *is horizontally regular if and only if* F_X *is absolutely continuous with respect to* $\sigma \otimes F_{X2}$ *and* $\int_{[-\pi,\pi)} \log \frac{dF_X(e^{i\lambda}, e^{i\theta})}{d(\sigma \otimes F_{X2})} \, d\sigma(e^{i\lambda}) > -\infty$ *for* $[F_{X2}]$-*a.e.* $e^{i\theta}$.

Proof. ($\Rightarrow$) We already established the fact that F_X is absolutely continuous with respect to $\sigma \otimes F_{X2}$ in Lemma 2.6.4. It remains to show that

$$\int_{[-\pi,\pi)} \log \frac{dF_X(e^{i\lambda}, e^{i\theta})}{d(\sigma \otimes F_{X2})} \, d\sigma(e^{i\lambda}) > -\infty \text{ for } [F_{X2}]\text{-a.e. } e^{i\theta}.$$

Recall that

$$\nu_{m,n}^1 = \int_{[-\pi,\pi)} \int_{[-\pi,\pi)} e^{im\lambda + in\theta} a(e^{i\lambda}, e^{i\theta}) \, d\mathcal{Z}_X(e^{i\lambda}, e^{i\theta}),$$

where $a(e^{i\lambda}, e^{i\theta}) = \mathcal{J}^{-1}(\nu_{0,0}^1)$. It then follows for $m \neq m'$ that

$$E\nu_{m,n}^1 \overline{\nu_{m',n'}^1} = \int_{[-\pi,\pi)} \int_{[-\pi,\pi)} e^{i(m-m')\lambda + i(n-n')\theta} |a(e^{i\lambda}, e^{i\theta})|^2 \, dF_X(e^{i\lambda}, e^{i\theta}) = 0.$$

Now employing the already established absolute continuity, we get

$$\int_{[-\pi,\pi)} \int_{[-\pi,\pi)} e^{i(m-m')\lambda + i(n-n')\theta} |a(e^{i\lambda}, e^{i\theta})|^2 \frac{dF_X(e^{i\lambda}, e^{i\theta})}{d(\sigma \otimes F_{X2})} \, d\sigma(e^{i\lambda}) dF_{X2}(e^{i\theta}) = 0.$$

Therefore, by the uniqueness of Fourier coefficients, we get that

$$\int_{[-\pi,\pi)} e^{i(m-m')\lambda} |a(e^{i\lambda}, e^{i\theta})|^2 \frac{dF_X(e^{i\lambda}, e^{i\theta})}{d(\sigma \otimes F_{X2})} \, d\sigma(e^{i\lambda}) = 0 \text{ for } [F_{X2}]\text{-a.e. } e^{i\theta}.$$

Since $m \neq m'$, if follows that for $[F_{X2}]$-a.e. $e^{i\theta}$,

$$|a(e^{i\lambda}, e^{i\theta})|^2 \frac{dF_X(e^{i\lambda}, e^{i\theta})}{d(\sigma \otimes F_{X2})} = C(e^{i\theta}) \text{ for } [\sigma]\text{-a.e. } e^{i\lambda}.$$

We point out that $C(e^{i\theta}) \neq 0$ since $E|\nu_{0,0}^1|^2 \neq 0$. We may then write for $[F_{X2}]$-a.e. $e^{i\theta}$,

$$\frac{dF_X(e^{i\lambda}, e^{i\theta})}{d(\sigma \otimes F_{X2})} = \frac{C(e^{i\theta})}{|a(e^{i\lambda}, e^{i\theta})|^2} \text{ for } [\sigma]\text{-a.e. } e^{i\lambda}. \tag{2.2}$$

We will conclude this direction of the proof by showing that for $[F_{X2}]$-a.e. $e^{i\theta}$, $\frac{1}{\overline{a(e^{i\lambda}, e^{i\theta})}} \in H^2(\mathbb{T}) = \overline{span}\{e^{in\lambda} : n \geq 0\}$ in $L^2(\mathbb{T}, \sigma)$. To this end, we have, for $m' < m$,

$$E\nu_{m,n}^1 \overline{X_{m',n'}} = \int_{[-\pi,\pi)} \int_{[-\pi,\pi)} e^{i(m-m')\lambda + i(n-n')\theta} a(\lambda, \theta) \, dF_X(e^{i\lambda}, e^{i\theta}) = 0.$$

Using absolute continuity and formula (2.2), we get for $[F_{X2}]$-a.e. $e^{i\theta}$ and $m' < m$,

$$C(e^{i\theta}) \int_{[-\pi,\pi)} e^{i(m-m')\lambda} \frac{1}{\overline{a}(e^{i\lambda}, e^{i\theta})} \, d\sigma(e^{i\lambda}) = 0.$$

It then follows that for $[F_{X2}]$-a.e. $e^{i\theta}$, $\dfrac{1}{\overline{a}(e^{i\lambda}, e^{i\theta})} \in H^2(\mathbb{T})$.

($\Leftarrow$) Under the given assumptions, it follows from the one parameter theory that for $[F_{X2}]$-a.e. $e^{i\theta}$, $\dfrac{dF_X(e^{i\lambda}, e^{i\theta})}{d(\sigma \otimes F_{X2})} = |b(e^{i\lambda}, e^{i\theta})|^2$, with $b(e^{i\lambda}, e^{i\theta}) \in H^2(\mathbb{T})$. Let $J : L(X) \to L^2(\sigma \otimes F_{X2})$ be the unitary operator that takes $X_{m,n}$ to $e^{im\lambda+in\theta}\overline{b}(e^{i\lambda}, e^{i\theta})$. Note that

$$J(L^1(X : m)) = \overline{span}\{e^{ik\lambda+in\theta}\overline{b}(e^{i\lambda}, e^{i\theta}) : k \leq m, n \in \mathbb{Z}\}.$$

It suffices to prove that $\cap_m J(L^1(X : m)) = \{0\}$. To see this, note that

$$\int_{[-\pi,\pi)} \int_{[-\pi,\pi)} e^{-ij\lambda+ik\lambda+in\theta}\overline{b}(e^{i\lambda}, e^{i\theta}) \, d\sigma(e^{i\lambda}) dF_{X2}(e^{i\theta}) = 0, \text{ for } k < j,$$

since $b(e^{i\lambda}, e^{i\theta}) \in H^2(\mathbb{T})$ for $[F_{X2}]$-a.e. $e^{i\theta}$. This says that

$$e^{ij\lambda} \perp \left\{e^{ik\lambda+in\theta}\overline{b}(e^{i\lambda}, e^{i\theta}) : k < j, n \in \mathbb{Z}\right\}.$$

Therefore, if $\phi \in \cap_m J(L^1(X : m))$ then $\phi \perp e^{ij\lambda}$ for all $j \in \mathbb{Z}$. Since $e^{in\theta}\phi \in \cap_m J(L^1(X : m))$, for all $n \in \mathbb{Z}$, it follows that $\phi \perp e^{ij\lambda+in\theta}$ for all $(j, n) \in \mathbb{Z}^2$. We can conclude from this that $\phi = 0$ $[\sigma \otimes F_{X2}]$-a.e. $\qquad\square$

This theorem allows us to construct the following family of $H^2(\mathbb{T})$ functions:

$$b(z, e^{i\theta}) = \exp\left[\frac{1}{2} \int_{[-\pi,\pi)} \frac{e^{i\lambda} + z}{e^{i\lambda} - z} \log\left(\frac{dF_X(e^{i\lambda}, e^{i\theta})}{d(\sigma \otimes F_{X2})}\right) d\sigma(e^{i\lambda})\right] \qquad |z| < 1.$$

It is well known from the theory of functions that for $[F_{X2}]$-a.e. $e^{i\theta}$,

$$\frac{dF_X(e^{i\lambda}, e^{i\theta})}{d(\sigma \otimes F_{X2})} = |b(e^{i\lambda}, e^{i\theta})|^2,$$

where $b(e^{i\lambda}, e^{i\theta})$ are the boundary values of $b(z, e^{i\theta})$.

Since $b(z, e^{i\theta})$ is a holomorphic function in the variable z, it may be written in the form

$$b(z, e^{i\theta}) = \sum_{k=0}^{\infty} \hat{b}_k(e^{i\theta})z^k,$$

and since

$$F_X\left(\{(e^{i\lambda}, e^{i\theta}) : b(e^{i\lambda}, e^{i\theta}) = 0 \; [\sigma \otimes F_{X2}]\text{-a.e.}\}\right) = 0,$$

we can define
$$Z_0(A) = \int_A \frac{1}{\overline{b}(e^{i\lambda}, e^{i\theta})} \, d\mathcal{Z}_X(e^{i\lambda}, e^{i\theta}).$$

Then,

$$
\begin{aligned}
X_{m,n} &= \int_{[-\pi,\pi)} \int_{[-\pi,\pi)} e^{im\lambda+in\theta} \, d\mathcal{Z}_X(e^{i\lambda}, e^{i\theta}) \\
&= \int_{[-\pi,\pi)} \int_{[-\pi,\pi)} e^{im\lambda+in\theta} \, \overline{b}(z, e^{i\theta}) \, dZ_0(e^{i\lambda}, e^{i\theta}) \\
&= \sum_{k=0}^{\infty} \int_{[-\pi,\pi)} \int_{[-\pi,\pi)} e^{im\lambda+in\theta} \, \overline{\hat{b}_k(e^{i\theta})} e^{-ik\lambda} \, dZ_0(e^{i\lambda}, e^{i\theta}).
\end{aligned}
$$

We observe that

$$P_{L^1(X:m-k)}X_{m,n} = \sum_{j=k}^{\infty} \int_{[-\pi,\pi)} \int_{[-\pi,\pi)} e^{im\lambda+in\theta} \, \overline{\hat{b}_j(e^{i\theta})} e^{-ij\lambda} \, dZ_0(e^{i\lambda}, e^{i\theta}),$$

and therefore,

$$
\begin{aligned}
&P_{L^1(X:m-k)\ominus L^1(X:m-k-1)}X_{m,n} \\
&= \int_{[-\pi,\pi)} \int_{[-\pi,\pi)} e^{im\lambda+in\theta} \, \overline{\hat{b}_k(e^{i\theta})} e^{-ik\lambda} \, dZ_0(e^{i\lambda}, e^{i\theta}).
\end{aligned}
$$

It then follows that

$$
\begin{aligned}
\nu^1_{m,n} &= \int_{[-\pi,\pi)} \int_{[-\pi,\pi)} e^{im\lambda+in\theta} \, \overline{\hat{b}_0(e^{i\theta})} \, dZ_0(e^{i\lambda}, e^{i\theta}) \\
&= \int_{[-\pi,\pi)} \int_{[-\pi,\pi)} e^{im\lambda+in\theta} \, \frac{\overline{\hat{b}_0(e^{i\theta})}}{\overline{b}(e^{i\lambda}, e^{i\theta})} \, d\mathcal{Z}_X(e^{i\lambda}, e^{i\theta}).
\end{aligned}
$$

It follows from this that $dF_{\nu^1}(e^{i\lambda}, e^{i\theta}) = |\hat{b}_0(e^{i\theta})|^2 \, d(\sigma \otimes F_{X2})(e^{i\lambda}, e^{i\theta})$. Therefore,
$$dF_{\nu^1 2}(e^{i\theta}) = |\hat{b}_0(e^{i\theta})|^2 \, dF_{X2}(e^{i\theta}).$$

We also point out that by our construction,

$$|\hat{b}_0(e^{i\theta})|^2 = \exp\left[\int_{[-\pi,\pi)} \log\left(\frac{dF_X(e^{i\lambda}, e^{i\theta})}{d(\sigma \otimes F_{X2})}\right) d\sigma(e^{i\lambda})\right].$$

Now, let $f_2 \in L^1(\mathbb{T}, \sigma)$ denote the absolutely continuous part of F_{X2} with respect to σ and let $f_{\nu^1,2} \in L^1(\mathbb{T}, \sigma)$ denote the absolutely continuous part of $F_{\nu^1 2}$ with respect to σ. So, we may then write

$$
\begin{aligned}
f_{\nu^1,2}(e^{i\theta}) &= \frac{dF_{\nu^1 2}(e^{i\theta})}{dF_{X2}} \cdot f_2(e^{i\theta}) = |\hat{b}_0(e^{i\theta})|^2 \cdot f_2(e^{i\theta}) \\
&= \exp\left[\int_{[-\pi,\pi)} \log\left(\frac{dF_X(e^{i\lambda}, e^{i\theta})}{d(\sigma \otimes F_{X2})}\right) d\sigma(e^{i\lambda})\right] \cdot f_2(e^{i\theta}) \\
&= \exp\left[\int_{[-\pi,\pi)} \log\left(\frac{dF_X(e^{i\lambda}, e^{i\theta})}{d(\sigma \otimes F_{X2})} \cdot f_2(e^{i\theta})\right) d\sigma(e^{i\lambda})\right].
\end{aligned}
$$

When $X_{m,n}$ is horizontally regular, we have that F_X is absolutely continuous with respect to $\sigma \otimes F_{X2}$. Therefore, letting $f_X \in L^1(\mathbb{T}^2, \sigma^2)$ denote the absolutely continuous part of F_X with respect to σ^2, we get

$$f_X(e^{i\lambda}, e^{i\theta}) = \frac{dF_X(e^{i\lambda}, e^{i\theta})}{d(\sigma \otimes F_{X2})} \cdot f_2(e^{i\theta}),$$

and so

$$f_{\nu^1,2}(e^{i\theta}) = \exp \left[\int_{[-\pi,\pi)} \log \left(f_X(e^{i\lambda}, e^{i\theta}) \right) d\sigma(e^{i\lambda}) \right]. \tag{2.3}$$

2.7 Spectral Measures and Spectral-type Wold Decompositions

We now examine the horizontally singular random fields. Let $g_a(e^{i\lambda}, e^{i\theta})$ denote the Radon-Nikodym derivative of the absolutely continuous part of F_X with respect to $\sigma \otimes F_{X2}$. Then, we have

Theorem 2.7.1. If $X_{m,n}$, $(m,n) \in \mathbb{Z}^2$ is a weakly stationary random field with $\int_{[-\pi,\pi)} \log g_a(e^{i\lambda}, e^{i\theta}) d\sigma(e^{i\lambda}) = -\infty$ for $[F_{X2}]$-a.e. $e^{i\theta}$, then $X_{m,n}$, $(m,n) \in \mathbb{Z}^2$ is horizontally singular.

Proof. Suppose the condition is satisfied, but $X_{m,n}$, $(m,n) \in \mathbb{Z}^2$ is not horizontally singular. Then, by the Horizontal Wold Decomposition,

$$X_{m,n} = X_{m,n}^r + X_{m,n}^s,$$

with $X_{m,n}^r \neq 0$. From this, we get that

$$F_X = F_{X^r} + F_{X^s},$$

where F_{X^p} is the spectral measure of X^p, for $p = r, s$. It follows that

$$F_{X2} = F_{X^r2} + F_{X^s2(2)}$$

and

$$g_a(e^{i\lambda}, e^{i\theta}) = g_a^r(e^{i\lambda}, e^{i\theta}) + g_a^s(e^{i\lambda}, e^{i\theta}),$$

where $g_a^p(e^{i\lambda}, e^{i\theta})$ is the density of the absolutely continuous part of F_{X^p} with respect to $\sigma \otimes F_{X2}$, for $p = r, s$. By Theorem 2.6.1, F_{X^r} is absolutely continuous with respect to $\sigma \otimes F_{X^r2}$ and by the equation above, we get that F_{X^r} is absolutely continuous with respect to $\sigma \otimes F_{X2}$. Therefore, for $A \in \mathcal{B}(\mathbb{T}^2)$, we get

$$F_{X^r}(A) = \iint_A g_a^r \, d(\sigma \otimes F_{X2}) = \iint_A g_a^r \, d(\sigma \otimes F_{X^r2}) + \iint_A g_a^r \, d(\sigma \otimes F_{X^s2}). \tag{2.4}$$

Let

$$F_{X^s 2}(B) = \int_B g \, dF_{X^r 2} + H(B), \qquad B \in \mathcal{B}(\mathbb{T}) \tag{2.5}$$

be the Lebesgue decomposition of $F_{X^s 2}$ with respect to $F_{X^r 2}$. From (2.4) and (2.5), we have

$$
\begin{aligned}
\frac{dF_{X^r}(e^{i\lambda}, e^{i\theta})}{d(\sigma \otimes F_{X^r 2})}
&= g_a^r(e^{i\lambda}, e^{i\theta}) + g_a^r(e^{i\lambda}, e^{i\theta}) \frac{d(\sigma \otimes F_{X^s 2})(e^{i\lambda}, e^{i\theta})}{d(\sigma \otimes F_{X^r 2})} \\
&= g_a^r(e^{i\lambda}, e^{i\theta}) + g_a^r(e^{i\lambda}, e^{i\theta}) g(e^{i\theta}) \\
&= g_a^r(e^{i\lambda}, e^{i\theta})(1 + g(e^{i\theta})) \qquad [\sigma \otimes F_{X2}]\text{-a.e.} \tag{2.6}
\end{aligned}
$$

Since X^r is horizontally regular, we get from Theorem 2.6.1 that

$$\int_{[-\pi,\pi)} \log \frac{dF_{X^r}(e^{i\lambda}, e^{i\theta})}{d(\sigma \otimes F_{X^r 2})} \, d\sigma(e^{i\lambda}) > -\infty \text{ for } [F_{X^r 2}]\text{-a.e. } e^{i\theta}.$$

It follows from (2.6) that $\displaystyle\int_{[-\pi,\pi)} \log g_a^r(e^{i\lambda}, e^{i\theta}) \, d\sigma(e^{i\lambda}) > -\infty$ for $[F_{X2}]$-a.e. $e^{i\theta}$, which contradicts the fact that $\displaystyle\int_{[-\pi,\pi)} \log g_a(e^{i\lambda}, e^{i\theta}) \, d\sigma(e^{i\lambda}) = -\infty$ for $[F_{X2}]$-a.e. $e^{i\theta}$. $\qquad\square$

Let $D = \left\{ e^{i\theta} : \displaystyle\int_{[-\pi,\pi)} \log g_a(e^{i\lambda}, e^{i\theta}) \, d\sigma(e^{i\lambda}) > -\infty \right\}$, and let $F_X = G + H$ be the Lebesgue decomposition of F_X with respect to $\sigma \otimes F_{X2}$. Let H be concentrated on the measurable set M. Let

$$N = M^c \cap (\mathbb{T} \times D)$$

and let $\mathcal{Z}_X$ be the orthogonally scattered set function corresponding to the spectral representation of $X_{m,n}$; that is,

$$X_{m,n} = \int_{[-\pi,\pi)} \int_{[-\pi,\pi)} e^{im\lambda + in\theta} \, d\mathcal{Z}_X(e^{i\lambda}, e^{i\theta}).$$

Now, define $Z^{(1)}(A) = \mathcal{Z}_X(A \cap N)$, $Z^{(2)}(A) = \mathcal{Z}_X(A \cap N^c)$,

$$X_{m,n}^{(1)} = \int_{[-\pi,\pi)} \int_{[-\pi,\pi)} e^{im\lambda + in\theta} \, dZ^{(1)}(e^{i\lambda}, e^{i\theta})$$

and

$$X_{m,n}^{(2)} = \int_{[-\pi,\pi)} \int_{[-\pi,\pi)} e^{im\lambda + in\theta} \, dZ^{(2)}(e^{i\lambda}, e^{i\theta}).$$

Henceforth, we will denote the spectral measure of $X^{(j)}$ by $F^{(j)}$, for $j = 1, 2$.

Theorem 2.7.2. $X_{m,n}^{(1)}$, $(m, n) \in \mathbb{Z}^2$ *is horizontally regular.*

Proof. By Theorem 2.6.1, we need to show that $F^{(1)}$ is absolutely continuous with respect to $\sigma \otimes F_2^{(1)}$ and

$$\int_{[-\pi,\pi)} \log \frac{dF^{(1)}(e^{i\lambda}, e^{i\theta})}{d(\sigma \otimes F_2^{(1)})} \, d\sigma(e^{i\lambda}) > -\infty \text{ for } [F_2^{(1)}]\text{-a.e. } e^{i\theta}.$$

Let $A \in \mathcal{B}(\mathbb{T}^2)$, then

$$
\begin{aligned}
F^{(1)}(A) &= \iint_A dF^{(1)} = \iint_{A \cap N} dF_X \\
&= \iint_{A \cap M^c \cap (\mathbb{T} \times D)} g_a \, d(\sigma \otimes F_{X2}) \\
&= \iint_{A \cap (\mathbb{T} \times D)} g_a \, d(\sigma \otimes F_{X2}) \\
&= \iint_A \mathbf{1}_{\mathbb{T} \times D} \, g_a \, d(\sigma \otimes F_{X2}).
\end{aligned}
$$

Let $B \in \mathcal{B}(\mathbb{T})$, then

$$
\begin{aligned}
F_2^{(1)}(B) &= F^{(1)}(\mathbb{T} \times B) = \iint_{\mathbb{T} \times B} \mathbf{1}_{\mathbb{T} \times D} \, g_a \, d(\sigma \otimes F_{X2}) \\
&= \int_B \left[\int_{[-\pi,\pi)} \mathbf{1}_{\mathbb{T} \times D}(e^{i\lambda}, \eta) g_a(e^{i\lambda}, \eta) \, d\sigma(e^{i\lambda}) \right] dF_{X2}(\eta) \\
&= \int_B g(\eta) \, dF_{X2}(\eta),
\end{aligned}
$$

where

$$g(\eta) = \begin{cases} \int_{[-\pi,\pi)} g_a(e^{i\lambda}, \eta) \, d\sigma(e^{i\lambda}) & \text{if } \eta \in D \\ 0 & \text{if } \eta \notin D. \end{cases}$$

Now, if we define

$$\ell(\xi, \eta) = \begin{cases} 1 & \text{if } g(\eta) = 0 \\ \frac{1}{g(\eta)} \mathbf{1}_{\mathbb{T} \times D}(\xi, \eta) g_a(\xi, \eta) & \text{if } g(\eta) > 0, \end{cases}$$

then

$$F^{(1)}(A) = \iint_A \ell(\xi, \eta) g(\eta) \, d(\sigma \otimes F_{X2}) = \iint_A \ell(\xi, \eta) \, d(\sigma \otimes F_2^{(1)}).$$

From this we see that $F^{(1)}$ is absolutely continuous with respect to $\sigma \otimes F_2^{(1)}$ with $\dfrac{dF^{(1)}(\xi, \eta)}{d(\sigma \otimes F_2^{(1)})} = \ell(\xi, \eta)$. A straightforward calculation shows that

$$\int_{[-\pi,\pi)} \log \frac{dF^{(1)}(e^{i\lambda}, e^{i\theta})}{d(\sigma \otimes F_2^{(1)})} \, d\sigma(e^{i\lambda}) > -\infty \text{ for } [F_2^{(1)}]\text{-a.e. } e^{i\theta},$$

as desired. $\qquad \square$

Continuing the ideas developed above. We point out that N^c can be written as the disjoint union $N^c = M \cup (M^c \cap (\mathbb{T} \times D^c))$. From this observation, we can further decompose $Z^{(2)}$ as follows. For $A \in \mathcal{B}(\mathbb{T}^2)$, we define

$$Z_s^{(2)}(A) = Z(A \cap M)$$

and

$$Z_a^{(2)}(A) = Z(A \cap (M^c \cap (\mathbb{T} \times D^c))).$$

From these definitions, we observe that

$$Z^{(2)}(A) = Z_a^{(2)}(A) + Z_s^{(2)}(A).$$

From this observation, we define

$$X_{m,n}^{(2)}(a) = \int_{[-\pi,\pi)} \int_{[-\pi,\pi)} e^{im\lambda + in\theta} \, dZ_a^{(2)}(e^{i\lambda}, e^{i\theta})$$

and

$$X_{m,n}^{(2)}(s) = \int_{[-\pi,\pi)} \int_{[-\pi,\pi)} e^{im\lambda + in\theta} \, dZ_s^{(2)}(e^{i\lambda}, e^{i\theta})$$

and observe that

$$X_{m,n}^{(2)} = X_{m,n}^{(2)}(a) + X_{m,n}^{(2)}(s).$$

Putting this together with our earlier decomposition, we get

$$X_{m,n} = X_{m,n}^{(1)} + X_{m,n}^{(2)}(a) + X_{m,n}^{(2)}(s), \text{ for all } (m,n) \in \mathbb{Z}^2.$$

By construction, these weakly stationary random fields are orthogonal and so,

$$L^1(X : m) = L^1(X^{(1)} : m) \oplus L^1(X^{(2)}(a) : m) \oplus L^1(X^{(2)}(s) : m), \text{ for all } m \in \mathbb{Z}.$$

Therefore, we get that

$$L^1(X : -\infty) = L^1(X^{(2)}(a) : -\infty) \oplus L^1(X^{(2)}(s) : -\infty),$$

since $X^{(1)}$ is horizontally regular by the above lemma. The next theorem follows from these observations.

Theorem 2.7.3. *If $X_{m,n}$, $(m,n) \in \mathbb{Z}^2$ is a weakly stationary random field that is horizontally singular, then* $\int_{[-\pi,\pi)} \log g_a(e^{i\lambda}, e^{i\theta}) \, d\sigma(e^{i\lambda}) = -\infty$ *for* $[F_{X2}]$-*a.e.* $e^{i\theta}$

We summarize some of our findings in the following theorem.

Theorem 2.7.4 (Horizontal Wold Decomposition (Spectral Form)). *Using the notation defined above, we get that a weakly stationary random field $X_{m,n}$, $(m,n) \in \mathbb{Z}^2$ has the following decomposition:*

$$X_{m,n} = X_{m,n}^{(1)} + X_{m,n}^{(2)}(a) + X_{m,n}^{(2)}(s), \text{ for all } (m,n) \in \mathbb{Z}^2,$$

with

    1. $L^1(X : m) = L^1(X^{(1)} : m) \oplus L^1(X^{(2)}(a) : m) \oplus L^1(X^{(2)}(s) : m)$, for all $m \in \mathbb{Z}$,

    2. $F_{X^{(1)}} = F_X(N \cap \cdot)$, $F_{X^{(2)}} = F_X(N^c \cap \cdot)$, where $X_{m,n}^{(2)} = X_{m,n}^{(2)}(a) + X_{m,n}^{(2)}(s)$,

    3. $X_{m,n}^{(1)}$ is horizontally regular, and

    4. $X_{m,n}^{(2)}$ is horizontally singular.

Next, we will work towards a Fourfold Wold Decomposition. To do this, we first need to set up some of the notation for the Vertical Wold Decomposition. Let $C = \left\{ e^{i\lambda} : \int_{[-\pi,\pi)} \log h_a(e^{i\lambda}, e^{i\theta}) \, d\sigma(e^{i\theta}) > -\infty \right\}$, where h_a denote the Radon-Nikodym derivative of the absolutely continuous part of F_X with respect to $F_{X1} \otimes \sigma$ and let $F_X = G + H$ be the Lebesgue decomposition of F_X with respect to $F_{X1} \otimes \sigma$. Let H be concentrated on the measurable set $\tilde{M}$. Let

$$\tilde{N} = \tilde{M}^c \cap (C \times \mathbb{T})$$

and let $\mathcal{Z}_X$ be the orthogonally scattered set function corresponding to the spectral representation of $X_{m,n}$; that is,

$$X_{m,n} = \int_{[-\pi,\pi)} \int_{[-\pi,\pi)} e^{im\lambda + in\theta} \, d\mathcal{Z}_X(e^{i\lambda}, e^{i\theta}).$$

Now, define $\tilde{Z}^{(1)}(A) = \mathcal{Z}_X(A \cap \tilde{N})$, $\tilde{Z}^{(2)}(A) = \mathcal{Z}_X(A \cap \tilde{N}^c)$,

$$\tilde{X}_{m,n}^{(1)} = \int_{[-\pi,\pi)} \int_{[-\pi,\pi)} e^{im\lambda + in\theta} \, d\tilde{Z}^{(1)}(e^{i\lambda}, e^{i\theta})$$

and

$$\tilde{X}_{m,n}^{(2)} = \int_{[-\pi,\pi)} \int_{[-\pi,\pi)} e^{im\lambda + in\theta} \, d\tilde{Z}^{(2)}(e^{i\lambda}, e^{i\theta}).$$

At this point, there is no reason to further decompose $\tilde{X}_{m,n}^{(2)}$. We now state the vertical analog of the spectral form of the Horizontal Wold Decomposition for the vertical case.

Theorem 2.7.5 (Vertical Wold Decomposition (Spectral Form)). *Using the* *notation defined above, we get that a weakly stationary random field* $X_{m,n}$, $(m,n) \in \mathbb{Z}^2$ *has the following decomposition:*

$$X_{m,n} = \tilde{X}^{(1)}_{m,n} + \tilde{X}^{(2)}_{m,n}, \text{ for all } (m,n) \in \mathbb{Z}^2,$$

with

1. $L^2(X:m) = L^2(\tilde{X}^{(1)}:m) \oplus L^2(\tilde{X}^{(2)}:m)$, *for all* $m \in \mathbb{Z}$,
2. $F_{\tilde{X}^{(1)}} = F_X(\tilde{N} \cap \cdot)$, $F_{\tilde{X}^{(2)}} = F_X(\tilde{N}^c \cap \cdot)$,
3. $\tilde{X}^{(1)}_{m,n}$ *is vertically regular, and*
4. $\tilde{X}^{(2)}_{m,n}$ *is vertically singular.*

One may then attempt to combine these decompositions in the following way. Observing that

$$\begin{aligned} \mathbb{T}^2 &= (N \cap N^c) \cup (\tilde{N} \cap \tilde{N}^c) \\ &= (N \cap \tilde{N}) \cup (N^c \cap \tilde{N}) \cap (N \cap \tilde{N}^c) \cup (N^c \cap \tilde{N}^c), \end{aligned}$$

we can define for each $A \in \mathcal{B}(\mathbb{T}^2)$,

$$\hat{Z}^{(1)}(A) = Z(A \cap (N \cap \tilde{N})),$$
$$\hat{Z}^{(2)}(A) = Z(A \cap (N^c \cap \tilde{N})),$$
$$\hat{Z}^{(3)}(A) = Z(A \cap (N \cap \tilde{N}^c)),$$

and

$$\hat{Z}^{(4)}(A) = Z(A \cap (N^c \cap \tilde{N}^c)).$$

We then define

$$\hat{X}^{(1)}_{m,n} = \int_{[-\pi,\pi)} \int_{[-\pi,\pi)} e^{im\lambda + in\theta} \, d\hat{Z}^{(1)}(e^{i\lambda}, e^{i\theta}),$$

$$\hat{X}^{(2)}_{m,n} = \int_{[-\pi,\pi)} \int_{[-\pi,\pi)} e^{im\lambda + in\theta} \, d\hat{Z}^{(2)}(e^{i\lambda}, e^{i\theta}),$$

$$\hat{X}^{(3)}_{m,n} = \int_{[-\pi,\pi)} \int_{[-\pi,\pi)} e^{im\lambda + in\theta} \, d\hat{Z}^{(3)}(e^{i\lambda}, e^{i\theta}),$$

and

$$\hat{X}^{(4)}_{m,n} = \int_{[-\pi,\pi)} \int_{[-\pi,\pi)} e^{im\lambda + in\theta} \, d\hat{Z}^{(4)}(e^{i\lambda}, e^{i\theta}).$$

By construction, we see that we get the following decomposition

$$X_{m,n} = \hat{X}^{(1)}_{m,n} + \hat{X}^{(2)}_{m,n} + \hat{X}^{(3)}_{m,n} + \hat{X}^{(4)}_{m,n}, \text{ for all } (m,n) \in \mathbb{Z}^2.$$

Although this is a fourfold orthogonal decomposition, it does not satisfy the desired conditions that we are looking for in such a decomposition, as you will see in the next example.

Example 2.7.1. *Let* $A = [0, \pi] \times [0, \pi]$, $B = [0, \pi] \times [\pi, 2\pi]$, *and* $C = [\pi, 2\pi] \times [0, \pi]$, *then define*

$$dF_X(e^{i\lambda}, e^{i\theta}) = \mathbf{1}_{A \cup B \cup C}(\lambda, \theta) \, d\sigma^2(e^{i\lambda}, e^{i\theta}).$$

It follows then that

$$dF_{X^{(1)}}(e^{i\lambda}, e^{i\theta}) = \mathbf{1}_{A \cup C}(\lambda, \theta) \, d\sigma^2(e^{i\lambda}, e^{i\theta}), \quad dF_{X^{(2)}}(e^{i\lambda}, e^{i\theta}) = \mathbf{1}_B(\lambda, \theta) \, d\sigma^2(e^{i\lambda}, e^{i\theta}),$$

$$dF_{\tilde{X}^{(1)}}(e^{i\lambda}, e^{i\theta}) = \mathbf{1}_{A \cup B}(\lambda, \theta) \, d\sigma^2(e^{i\lambda}, e^{i\theta}), \quad dF_{\tilde{X}^{(2)}}(e^{i\lambda}, e^{i\theta}) = \mathbf{1}_C(\lambda, \theta) \, d\sigma^2(e^{i\lambda}, e^{i\theta}),$$

$$dF_{\hat{X}^{(1)}}(e^{i\lambda}, e^{i\theta}) = \mathbf{1}_A(\lambda, \theta) \, d\sigma^2(e^{i\lambda}, e^{i\theta}), \quad dF_{\hat{X}^{(2)}}(e^{i\lambda}, e^{i\theta}) = \mathbf{1}_B(\lambda, \theta) \, d\sigma^2(e^{i\lambda}, e^{i\theta}),$$

$$dF_{\hat{X}^{(3)}}(e^{i\lambda}, e^{i\theta}) = \mathbf{1}_C(\lambda, \theta) \, d\sigma^2(e^{i\lambda}, e^{i\theta}), \quad \text{and} \quad dF_{\hat{X}^{(4)}}(e^{i\lambda}, e^{i\theta}) = 0 \, d\sigma^2(e^{i\lambda}, e^{i\theta}).$$

If this gave a true fourfold Wold Decomposition, one would expect that $\hat{X}^{(1)}$ *would be both horizontally regular and vertically regular. On the contrary, we see from our above work that* $\hat{X}^{(1)}$ *is both horizontally singular and vertically singular. This example shows that additional conditions are needed on the spectral measure to get the desired fourfold decomposition.*

We now want to obtain conditions on F_X to get a fourfold Wold Decomposition. Let

$$dF_X(e^{i\lambda}, e^{i\theta}) = f_a(e^{i\lambda}, e^{i\theta}) \, d\sigma^2(e^{i\lambda}, e^{i\theta}) + dF^s(e^{i\lambda}, e^{i\theta})$$

be the Lebesgue decomposition of F_X with respect to σ^2 and let Δ^s be the support of F^s and $\Delta^a = (\Delta^s)^c$. Then, as before, we get

$$X_{m,n} = X_{m,n}^a + X_{m,n}^s, \tag{2.7}$$

where

$$X_{m,n}^a = \iint_{\Delta^a} e^{im\lambda + in\theta} \, d\mathcal{Z}_X(e^{i\lambda}, e^{i\theta}),$$

and

$$X_{m,n}^s = \iint_{\Delta^s} e^{im\lambda + in\theta} \, d\mathcal{Z}_X(e^{i\lambda}, e^{i\theta}).$$

These are mutually orthogonal stationary random fields and

$$L^j(X : m) = L^j(X^a : m) \oplus L^j(X^s : m), \text{ for all } m \in \mathbb{Z}, \ j = 1, 2.$$

We now get the following theorem.

Theorem 2.7.6. *The random field* $X_{m,n}^s$, $(m, n) \in \mathbb{Z}^2$ *admits the following unique decomposition*

$$X_{m,n}^s = X_{m,n}^{r,s} + X_{m,n}^{s,r} + X_{m,n}^{s,s},$$

where

1. $X_{m,n}^{r,s}$, $X_{m,n}^{s,r}$, and $X_{m,n}^{s,s}$ are mutually orthogonal random fields.

2. $X_{m,n}^{r,s}$ is horizontally regular and vertically singular.

3. $X_{m,n}^{s,r}$ is horizontally singular and vertically regular.

4. $X_{m,n}^{s,s}$ is horizontally singular and vertically singular.

5. $L^j(X^s : m) = L^j(X^{r,s} : m) \oplus L^j(X^{s,r} : m) \oplus L^j(X^{s,s} : m)$, for all $m \in \mathbb{Z}$, $j = 1, 2$.

Proof. Using the spectral form of the horizontal Wold Decompositions, we get that

$$X_{m,n}^s = X_{m,n}^{s,(1)} + X_{m,n}^{s,(2)},$$

with $F_{X^{s,(1)}}$ absolutely continuous with respect to $\sigma \otimes F_2^s$. Similarly, using the spectral form of the vertical Wold Decompositions, we get that

$$X_{m,n}^s = \tilde{X}_{m,n}^{s,(1)} + \tilde{X}_{m,n}^{s,(2)},$$

with $F_{\tilde{X}^{s,(1)}}$ absolutely continuous with respect to $F_1^s \otimes \sigma$. It then follows that $X_{m,n}^{s,(1)}$ is orthogonal to $\tilde{X}_{m,n}^{s,(1)}$. Again, using the Wold Decomposition in each direction, we have that

$$L(X^s) = L(X^{s,(1)}) \oplus L^1(X^s : -\infty)$$

and

$$L(X^s) = L(\tilde{X}^{s,(1)}) \oplus L^2(X^s : -\infty).$$

These decompositions and the remark above imply that

$$L(X^{s,(1)}) \subseteq L^2(X^s : -\infty)$$

and

$$L(\tilde{X}^{s,(1)}) \subseteq L^1(X^s : -\infty).$$

It then follows that $X^{s,(1)}$ is both horizontally regular and vertically singular, and that $\tilde{X}^{s,(1)}$ is both horizontally singular and vertically regular. Further, we have that

$$
\begin{aligned}
L(X^s) &= L(X^{s,(1)}) \oplus L(\tilde{X}^{s,(1)}) \oplus \left(L^1(X^s : -\infty) \ominus L(\tilde{X}^{s,(1)}) \right) \\
&= L(X^{s,(1)}) \oplus L(\tilde{X}^{s,(1)}) \oplus \left(L^1(X^s : -\infty) \cap L(\tilde{X}^{s,(2)}) \right) \\
&= L(X^{s,(1)}) \oplus L(\tilde{X}^{s,(1)}) \oplus \tilde{L}(X^s : -\infty),
\end{aligned}
$$

where $\tilde{L}(X^s : -\infty) = L^1(X^s : -\infty) \cap L^2(X^s : -\infty)$. We can see from this decomposition and our observations above that

$$X_{m,n}^s = X_{m,n}^{r,s} + X_{m,n}^{s,r} + X_{m,n}^{s,s},$$

where

1. $X_{m,n}^{r,s}$, $X_{m,n}^{s,r}$, and $X_{m,n}^{s,s}$ are mutually orthogonal random fields.

2. $X_{m,n}^{r,s} = P_{L^2(X^s:-\infty)\ominus\tilde{L}(X^s:-\infty)}X_{m,n}^s$ is horizontally regular and vertically singular.

3. $X_{m,n}^{s,r} = P_{L^1(X^s:-\infty)\ominus\tilde{L}(X^s:-\infty)}X_{m,n}^s$ is horizontally singular and vertically regular.

4. $X_{m,n}^{s,s} = P_{\tilde{L}(X^s:-\infty)}X_{m,n}^s$ is horizontally singular and vertically singular.

5. $L^j(X^s:m) = L^j(X^{r,s}:m) \oplus L^j(X^{s,r}:m) \oplus L^j(X^{s,s}:m)$, for all $m \in \mathbb{Z}$, $j = 1, 2$.

□

We see from this theorem that the absolutely continuous portion of our spectral measure is the part of the spectral measure that determines whether or not we are able to get a fourfold Wold Decomposition. The following theorem sheds light on the situation.

Theorem 2.7.7. *A stationary random field $X_{m,n}$, $(m,n) \in \mathbb{Z}^2$ admits the following unique fourfold decomposition*

$$X_{m,n} = X_{m,n}^{r,r} + X_{m,n}^{r,s} + X_{m,n}^{s,r} + X_{m,n}^{s,s},$$

where

1. $X_{m,n}^{r,r}$, $X_{m,n}^{r,s}$, $X_{m,n}^{s,r}$, and $X_{m,n}^{s,s}$ *are mutually orthogonal random fields.*

2. $X_{m,n}^{r,r}$ *is horizontally regular and vertically regular.*

3. $X_{m,n}^{r,s}$ *is horizontally regular and vertically singular.*

4. $X_{m,n}^{s,r}$ *is horizontally singular and vertically regular.*

5. $X_{m,n}^{s,s}$ *is horizontally singular and vertically singular.*

6. $L^j(X:m) = L^j(X^{r,r}:m) \oplus L^j(X^{r,s}:m) \oplus L^j(X^{s,r}:m) \oplus L^j(X^{s,s}:m)$, *for all $m \in \mathbb{Z}$, $j = 1, 2$.*

if and only if one of the following two conditions is satisfied.

1. $\displaystyle\int_{[-\pi,\pi)}\int_{[-\pi,\pi)} \log\left(\frac{dF_X(e^{i\lambda}, e^{i\theta})}{d\sigma^2}\right) d\sigma^2(e^{i\lambda}, e^{i\theta}) > -\infty$

2. *The random field $X_{m,n}^a$, $(m,n) \in \mathbb{Z}^2$ is horizontally singular or vertically singular.*

A majority of this theorem follows from the work we have done above. However, further analysis is needed. The following are some preliminary results needed to prove this theorem. Our first result is a variation of Theorem 2.6.1 and the proof is very similar. For this reason, we will not include a proof.

Theorem 2.7.8. *Let $X_{m,n}$, $(m,n) \in \mathbb{Z}^2$ be a not horizontally singular weakly stationary random field for which there exists a positive bounded measure μ so that F_X is absolutely continuous with respect to $\sigma \otimes \mu$, with $\dfrac{dF_X}{d(\sigma \otimes \mu)} = f_\mu$. Then,*

1. F_X is absolutely continuous with respect to $\sigma \otimes F_2$, and F_{ν_1} is absolutely continuous with respect to $\sigma \otimes \mu$.

2. $X_{m,n}$, $(m,n) \in \mathbb{Z}^2$ is horizontally regular if and only if

$$\int_{[-\pi,\pi)} \log f_\mu(e^{i\lambda}, e^{i\theta}) \, d\sigma(e^{i\lambda}) > -\infty$$

on $B_\mu = \left\{ e^{i\theta} : \displaystyle\int_{[-\pi,\pi)} f_\mu(e^{i\lambda}, e^{i\theta}) \, d\sigma(e^{i\lambda}) > 0 \right\}$.

3. $A_\mu = \left\{ e^{i\theta} : \displaystyle\int_{[-\pi,\pi)} \log f_\mu(e^{i\lambda}, e^{i\theta}) \, d\sigma(e^{i\lambda}) > -\infty \right\}$ has positive μ measure.

In the proof of the next lemma, we will need the following notation, let $A_\mu^\theta = \{\theta \in [-\pi, \pi) : e^{i\theta} \in A_\mu\}$.

Lemma 2.7.1. *If $dF_X(e^{i\lambda}, e^{i\theta}) = f(e^{i\lambda}, e^{i\theta}) \, d\sigma^2(e^{i\lambda}, e^{i\theta})$, then $X_{m,n}$, $(m,n) \in \mathbb{Z}^2$ is not horizontally singular and not vertically singular if and only if $[L(X) \ominus L^1(X : -\infty)] \cap [L(X) \ominus L^2(X : -\infty)] \neq \{0\}$.*

Proof. Sufficiency is clear. To get necessity observe that the horizontally regular part of $X_{m,n}$ is given by $\displaystyle\int_{[-\pi,\pi)} \int_{A_\sigma^\theta} e^{im\lambda + in\theta} \, dZ_X(e^{i\lambda}, e^{i\theta})$ and the vertically regular part is given by $\displaystyle\int_{A_\sigma^\lambda} \int_{[-\pi,\pi)} e^{im\lambda + in\theta} \, dZ_X(e^{i\lambda}, e^{i\theta})$, where $A_\sigma^\lambda = \left\{ \lambda \in [-\pi, \pi) : \displaystyle\int_{[-\pi,\pi)} \log f(e^{i\lambda}, e^{i\theta}) \, d\sigma(e^{i\theta}) > -\infty \right\}$. By the above theorem, A_σ and $A_\sigma' = \{e^{i\lambda} : \lambda \in A_\sigma^\lambda\}$ are not of Lebesgue measure zero. This shows that the projection of $X_{m,n}$ on the intersection of $[L(X) \ominus L^1(X : -\infty)] \cap [L(X) \ominus L^2(X : -\infty)]$ is nonzero. $\square$

Theorem 2.7.9. *If $dF_X(e^{i\lambda}, e^{i\theta}) = f(e^{i\lambda}, e^{i\theta}) \, d\sigma^2(e^{i\lambda}, e^{i\theta})$, then $X_{m,n}$, $(m,n) \in \mathbb{Z}^2$ is horizontally regular and vertically regular if and only if*

1. $\displaystyle\int_{[-\pi,\pi)} \log f(e^{i\lambda}, e^{i\theta}) \, d\sigma(e^{i\lambda}) > -\infty$ for $[\sigma]$-a.e. $e^{i\theta}$, and

2. $\displaystyle\int_{[-\pi,\pi)} \log f(e^{i\lambda}, e^{i\theta}) \, d\sigma(e^{i\theta}) > -\infty$ for $[\sigma]$-a.e. $e^{i\lambda}$.

Proof. Let $B_1 = \left\{ e^{i\theta} : \int_{[-\pi,\pi)} f(e^{i\lambda}, e^{i\theta}) \, d\sigma(e^{i\lambda}) > 0 \right\}$ and

$B_2 = \left\{ e^{i\lambda} : \int_{[-\pi,\pi)} f(e^{i\lambda}, e^{i\theta}) \, d\sigma(e^{i\theta}) > 0 \right\}$. Then, we have from Theorem

2.7.8 that $X_{m,n}$, $(m,n) \in \mathbb{Z}^2$ is horizontally regular if and only if

$$\int_{[-\pi,\pi)} \log f(e^{i\lambda}, e^{i\theta}) \, d\sigma(e^{i\lambda}) > -\infty \text{ on } B_1,$$

which is true if and only if $\sigma(B_2^c) = 0$, which follows from Jensen's Inequality for concave functions. This then gives us that $X_{m,n}$, $(m,n) \in \mathbb{Z}^2$ is vertically regular if and only if $\int_{[-\pi,\pi)} \log f(e^{i\lambda}, e^{i\theta}) \, d\sigma(e^{i\theta}) > -\infty$ for $[\sigma]$-a.e. $e^{i\lambda}$. Now, interchanging the roles of horizontal regular and vertical regular in the above argument gives us our desired result. $\qquad\square$

Corollary 2.7.1. *If* $dF_X(e^{i\lambda}, e^{i\theta}) = f(e^{i\lambda}, e^{i\theta}) \, d\sigma^2(e^{i\lambda}, e^{i\theta})$ *and* $X_{m,n}$, $(m,n) \in \mathbb{Z}^2$ *is horizontally regular and vertically regular, then* $f(e^{i\lambda}, e^{i\theta}) > 0$ $[\sigma^2]$*-a.e.*

Corollary 2.7.2. *If* $dF_X(e^{i\lambda}, e^{i\theta}) = f(e^{i\lambda}, e^{i\theta}) \, d\sigma^2(e^{i\lambda}, e^{i\theta})$ *and* $X_{m,n}$, $(m,n) \in \mathbb{Z}^2$ *is horizontally regular and vertically regular, then*

$$X_{m,n} = P_{[L(X) \ominus L^1(X:-\infty)] \cap [L(X) \ominus L^2(X:-\infty)]} X_{m,n}$$

$$= \iint_{B_2^\lambda \times B_1^\theta} e^{im\lambda + in\theta} \, d\mathcal{Z}_X(e^{i\lambda}, e^{i\theta}),$$

where $B_1^\theta = \{\theta \in [-\pi,\pi) : e^{i\theta} \in B_1\}$ *and* $B_2^\lambda = \{\lambda \in [-\pi,\pi) : e^{i\lambda} \in B_2\}$*, with* B_1 *and* B_2 *being defined in the proof of the last theorem.*

At this point, we have developed enough theory to prove Theorem 2.7.7. We start by decomposing our random field as we did in Equation (2.7). Further, we decompose $X_{m,n}^s$ as we did in Theorem 2.7.6. If we set $X_{m,n}^{r,r} = X_{m,n}^a$, this gives us parts 1, 3, 4, 5, and 6. Then, using Lemma 2.7.1 and Theorem 2.7.9 and the fact that $X_{m,n}^{r,r} = P_{[L(X) \ominus L^1(X:-\infty)] \cap [L(X) \ominus L^2(X:-\infty)]} X_{m,n}$, since $X_{m,n}^{r,s} = P_{L^2(X:-\infty) \ominus \tilde{L}(X:-\infty)} X_{m,n}$, $X_{m,n}^{s,r} = P_{L^1(X:-\infty) \ominus \tilde{L}(X:-\infty)} X_{m,n}$, and $X_{m,n}^{s,s} = P_{\tilde{L}(X:-\infty)} X_{m,n}$, we can conclude that our theorem holds.

2.8 The Fourfold Wold Decomposition

Let $X_{m,n}$, $(m,n) \in \mathbb{Z}^2$ be a weakly stationary random field. We define the following subspaces of $L^2(\Omega, \mathcal{F}, P)$.

$$L(X) = \overline{span}\{X_{m,n} : (m,n) \in \mathbb{Z}^2\},$$

$$L(X:m,n) = \overline{span}\{X_{j,k} : j \le m, k \le n\},$$
$$L^1(X:m) = \overline{span}\{X_{j,k} : j \le m, k \in \mathbb{Z}\},$$
$$L^2(X:n) = \overline{span}\{X_{j,k} : j \in \mathbb{Z}, k \le n\},$$
$$L^1(X:-\infty) = \bigcap_m L^1(X:m),$$

and

$$L^2(X:-\infty) = \bigcap_n L^2(X:n).$$

The following condition will be used in this section. We will call it the **strong commuting condition**:

$$P_{L^1(X:m)} P_{L^2(X:n)} = P_{L(X:m,n)}, \quad \forall (m,n) \in \mathbb{Z}^2. \tag{2.8}$$

Note that the strong commuting condition implies what we will call the **weak commuting condition**:

$$P_{L^1(X:m)} P_{L^2(X:n)} = P_{L^2(X:n)} P_{L^1(X:m)}, \quad \forall (m,n) \in \mathbb{Z}^2. \tag{2.9}$$

This follows from the fact that $P_{L^1(X:m)} P_{L^2(X:n)} = P_{L(X:m,n)} = \left(P_{L(X:m,n)}\right)^* = \left(P_{L^1(X:m)} P_{L^2(X:n)}\right)^* = P^*_{L^2(X:n)} P^*_{L^1(X:m)} = P_{L^2(X:n)} P_{L^1(X:m)}$.

Theorem 2.8.1 (Wold (Weak) Fourfold Decomposition). *Let $X_{m,n}$, $(m,n) \in \mathbb{Z}^2$ be a weakly stationary random field satisfying (2.9), the weak commuting condition. Then, for each $(m,n) \in \mathbb{Z}^2$, we get*

$$L^1(X:m) \cap L^2(X:n)$$

$$= \left(\sum_{j \le m, k \le n} \oplus W_{j,k}\right) \oplus \left(\sum_{j \le m} \oplus I_j^1\right) \oplus \left(\sum_{k \le n} \oplus I_k^2\right) \oplus L(X:-\infty), \tag{2.10}$$

where

1. $W_{j,k} = \left[L^1(X:j) \ominus L^1(X:j-1)\right] \cap \left[L^2(X:k) \ominus L^2(X:k-1)\right]$
2. $I_j^1 = \left[L^1(X:j) \ominus L^1(X:j-1)\right] \cap L^2(X:-\infty)$
3. $I_k^2 = L^1(X:-\infty) \cap \left[L^2(X:k) \ominus L^2(X:k-1)\right]$
4. $L(X:-\infty) = L^1(X:-\infty) \cap L^2(X:-\infty)$.

Proof. For simplicity of notation, we define $W_j^1 = L^1(X:j) \ominus L^1(X:j-1)$ and $W_k^2 = L^2(X:k) \ominus L^2(X:k-1)$. It then follows from (2.9), the weak commuting condition, that $P_{W_j^1} P_{W_k^2} = P_{W_k^2} P_{W_j^1}$, $P_{W_j^1} P_{L^2(X:-\infty)} = P_{L^2(X:-\infty)} P_{W_j^1}$, $P_{W_k^2} P_{L^1(X:-\infty)} = P_{L^1(X:-\infty)} P_{W_k^2}$, and

$P_{L^1(X:-\infty)}P_{L^2(X:-\infty)} = P_{L^2(X:-\infty)}P_{L^1(X:-\infty)}$. It follows directly from the Wold Decomposition for stationary sequences that

$$P_{L^1(X:m)} = \sum_{j=-\infty}^{m} P_{W_j^1} + P_{L^1(X:-\infty)}$$

and

$$P_{L^2(X:n)} = \sum_{k=-\infty}^{n} P_{W_k^2} + P_{L^2(X:-\infty)}.$$

Multiplying these, we get the following two equations:

$$P_{L^1(X:m)}P_{L^2(X:n)} = \sum_{k=-\infty}^{n}\sum_{j=-\infty}^{m} P_{W_j^1}P_{W_k^2} + \sum_{k=-\infty}^{n} P_{L^1(X:-\infty)}P_{W_k^2}$$
$$+ \sum_{j=-\infty}^{m} P_{W_j^1}P_{L^2(X:-\infty)} + P_{L^1(X:-\infty)}P_{L^2(X:-\infty)}$$

and

$$P_{L^2(X:n)}P_{L^1(X:m)} = \sum_{k=-\infty}^{n}\sum_{j=-\infty}^{m} P_{W_k^2}P_{W_j^1} + \sum_{k=-\infty}^{n} P_{W_k^2}P_{L^1(X:-\infty)}$$
$$+ \sum_{j=-\infty}^{m} P_{L^2(X:-\infty)}P_{W_j^1} + P_{L^2(X:-\infty)}P_{L^1(X:-\infty)}.$$

Once again, from the weak commuting condition, the observations made above and a basis property of commuting projections (see the appendix), we have that

$$P_{L^1(X:m)\cap L^2(X:n)} = \sum_{k=-\infty}^{n}\sum_{j=-\infty}^{m} P_{W_j^1\cap W_k^2} + \sum_{j=-\infty}^{m} P_{W_j^1\cap L^2(X:-\infty)}$$
$$+ \sum_{k=-\infty}^{n} P_{L^1(X:-\infty)\cap W_k^2} + P_{L^1(X:-\infty)\cap L^2(X:-\infty)}.$$

$$(2.11)$$

Therefore,

$$L^1(X:m) \cap L^2(X:n)$$

$$= \left(\sum_{j\le m, k\le n} \oplus W_{j,k}\right) \oplus \left(\sum_{j\le m} \oplus I_j^1\right) \oplus \left(\sum_{k\le n} \oplus I_k^2\right) \oplus L(X:-\infty),$$

$$(2.12)$$

where

1. $W_{j,k} = \left[L^1(X:j) \ominus L^1(X:j-1)\right] \cap \left[L^2(X:k) \ominus L^2(X:k-1)\right]$,
2. $I_j^1 = \left[L^1(X:j) \ominus L^1(X:j-1)\right] \cap L^2(X:-\infty)$,
3. $I_k^2 = L^1(X:-\infty) \cap \left[L^2(X:k) \ominus L^2(X:k-1)\right]$, and
4. $L(X:-\infty) = L^1(X:-\infty) \cap L^2(X:-\infty)$,

as desired. $\qquad\qquad\square$

Theorem 2.8.2 (Wold (Strong) Fourfold Decomposition). *Let* $X_{m,n}$, $(m,n) \in \mathbb{Z}^2$ *be a weakly stationary random field satisfying (2.8), the strong commuting condition. Then, for each* $(m,n) \in \mathbb{Z}^2$, *we get*

$$L(X:m,n) = \left(\sum_{j \le m, k \le n} \oplus W_{j,k}\right) \oplus \left(\sum_{j \le m} \oplus I_j^1\right) \oplus \left(\sum_{k \le n} \oplus I_k^2\right) \oplus L(X:-\infty),$$

(2.13)

where

1. $W_{j,k} = \left[L^1(X:j) \ominus L^1(X:j-1)\right] \cap \left[L^2(X:k) \ominus L^2(X:k-1)\right]$
2. $I_j^1 = \left[L^1(X:j) \ominus L^1(X:j-1)\right] \cap L^2(X:-\infty)$
3. $I_k^2 = L^1(X:-\infty) \cap \left[L^2(X:k) \ominus L^2(X:k-1)\right]$
4. $L(X:-\infty) = \bigcap_{m,n} L(X:m,n) = L^1(X:-\infty) \cap L^2(X:-\infty)$.

Proof. It follows from (2.8), the strong commuting condition, and Equation (2.11) that

$$P_{L(X:m,n)} = \sum_{k=-\infty}^{n} \sum_{j=-\infty}^{m} P_{W_j^1 \cap W_k^2} + \sum_{j=-\infty}^{m} P_{W_j^1 \cap L^2(X:-\infty)}$$
$$+ \sum_{k=-\infty}^{n} P_{L^1(X:-\infty) \cap W_k^2} + P_{L^1(X:-\infty) \cap L^2(X:-\infty)}.$$

The theorem follows. $\qquad\qquad\square$

From Theorem 2.8.1, we get the following decomposition of the random field. Before we state the theorem, we introduce some new notation. Let

$$N_0 = \sum_{j,k} \oplus W_{j,k},$$

$$N_1 = \sum_{j} \oplus I_j^1,$$

and

$$N_2 = \sum_{k} \oplus I_k^2.$$

Theorem 2.8.3. *Let $X_{m,n}$, $(m,n) \in \mathbb{Z}^2$ be a weakly stationary random field satisfying (2.9), the weak commuting condition. Then, $X_{m,n}$ has the following unique decomposition into mutually orthogonal weakly stationary random fields:*

$$X_{m,n} = X_{m,n}^{r,r} + X_{m,n}^{r,s} + X_{m,n}^{s,r} + X_{m,n}^{s,s}, \tag{2.14}$$

where

1. *$X_{m,n}^{r,r} = P_{N_0} X_{m,n}$ is horizontally regular and vertically regular,*

2. *$X_{m,n}^{r,s} = P_{N_1} X_{m,n}$ is horizontally regular and vertically singular,*

3. *$X_{m,n}^{s,r} = P_{N_2} X_{m,n}$ is horizontally singular and vertically regular, and*

4. *$X_{m,n}^{s,s} = P_{L(X:-\infty)} X_{m,n}$ is horizontally singular and vertically singular, where $L(X : -\infty) = L^1(X : -\infty) \bigcap L^2(X : -\infty)$ as defined in Theorem 2.8.1.*

Furthermore, each of these weakly stationary random fields satisfies (2.9), the weak commuting condition.

Proof. By Theorem 2.8.1, $L(X) = N_0 \oplus N_1 \oplus N_2 \oplus L(X : -\infty)$. Therefore, decomposition (2.14) is mutually orthogonal. Since N_0, N_1, N_2 and $L(X : -\infty)$ are invariant under U_1 and U_2, it follows that $X_{m,n}^{r,r}$, $X_{m,n}^{r,s}$, $X_{m,n}^{s,r}$ and $X_{m,n}^{s,s}$ are weakly stationary random fields.

To see that $X_{m,n}^{r,r}$ is horizontally regular and vertically regular, note that by Lemma A.1.1, in the appendix, it follows that $P_{L^j(X^{r,r}:m)} = P_{N_0} P_{L^j(X:m)}$. Therefore, $P_{L^j(X^{r,r}:-\infty)} = P_{N_0} P_{L^j(X:-\infty)} = P_{N_0 \cap L^j(X:-\infty)} = P_{\{0\}}$. It follows that $L^j(X^{r,r} : -\infty) = \{0\}$ for $j = 1,2$. That is, $X_{m,n}^{r,r}$ is horizontally regular and vertically regular.

To see that $X_{m,n}^{r,s}$ is horizontally regular and vertically singular, note again by Lemma A.1.1 that $P_{L^j(X^{r,s}:m)} = P_{N_1} P_{L^j(X:m)}$ for $j = 1,2$. Therefore, $P_{L^1(X^{r,s}:-\infty)} = P_{N_1} P_{L^1(X:-\infty)} = P_{N_1 \cap L^1(X:-\infty)} = P_{\{0\}}$, $P_{L^2(X^{r,s}:-\infty)} = P_{N_1} P_{L^2(X:-\infty)} = P_{N_1 \cap L^2(X:-\infty)} = P_{N_1} = P_{N_1 \cap L(X)} = P_{N_1} P_{L(X)} = P_{L(X^{r,s})}$. It follows that $L^1(X^{r,s} : -\infty) = \{0\}$ and that $L^2(X^{r,s} : -\infty) = L(X^{r,s})$. That is, $X_{m,n}^{r,r}$ is horizontally regular and vertically singular.

A similar argument shows that $X_{m,n}^{s,r} = P_{N_2} X_{m,n}$ is horizontally singular and vertically regular.

Finally, to see that $X_{m,n}^{s,s} = P_{L(X:-\infty)} X_{m,n}$ is horizontally singular and vertically singular, observe once again from Lemma A.1.1 that $P_{L^j(X^{s,s}:m)} = P_{L(X:-\infty)} P_{L^j(X:m)}$ for $j = 1,2$. Therefore, we get $P_{L^j(X^{s,s}:-\infty)} = P_{L(X:-\infty)} P_{L^j(X:-\infty)} = P_{L(X:-\infty)} = P_{L(X:-\infty) \cap L(X)} = P_{L(X:-\infty)} P_{L(X)}$ for $j = 1,2$. Hence, $L^j(X^{s,s} : -\infty) = L(X^{s,s})$ for $j = 1,2$. That is, $X_{m,n}^{s,s} = P_{L(X:-\infty)} X_{m,n}$ is horizontally singular and vertically singular.

Once again, it follows from the method above using Lemma A.1.1 that $X_{m,n}^{r,r}$, $X_{m,n}^{r,s}$, $X_{m,n}^{s,r}$ and $X_{m,n}^{s,s}$ satisfy (2.9), the weak commuting condition. $\square$

We now take a more detailed look at some of the random fields from the last theorem.

Lemma 2.8.1. $P_{N_0} X_{m,n} = \sum_{j \leq m} \sum_{k \leq n} \sum_{l=1}^{M_0} a_0^{(l)}(m-j, n-k) \nu_{j,k}^{(l)}, \; where$

$a_0^{(l)}(j,k) \in \mathbb{C}$ *for all* j,k,l *with* $\sum_j \sum_k \sum_{l=1}^{M_0} |a_0^{(l)}(j,k)|^2 < \infty$, *and* $\nu_{j,k}^{(l)} =$

$U_1^j U_2^k \nu_{0,0}^{(l)}$, *where* $\left\{\nu_{0,0}^{(l)}\right\}_{1 \leq l \leq M_0}$ *is an orthonormal basis for* $W_{0,0}$.

Note: M_0 is clearly the dimension of $W_{0,0}$ and as such will satisfy the inequality $1 \leq M_0 \leq \infty$.

Proof. Let $\left\{\nu_{0,0}^{(l)}\right\}_{1 \leq l \leq M_0}$ is an orthonormal basis for $W_{0,0}$. Then, $\left\{\nu_{j,k}^{(l)}\right\}_{1 \leq l \leq M_0}$

$= \left\{U_1^j U_2^k \nu_{0,0}^{(l)}\right\}_{1 \leq l \leq M_0}$ is an orthonormal basis for $W_{j,k}$. Therefore,

$$P_{N_0} X_{m,n} = \sum_{j \leq m} \sum_{k \leq n} \sum_{l=1}^{M_0} a_0^{(l)}(j,k,m,n) \nu_{j,k}^{(l)}, \qquad '$$

where $a_0^{(l)}(j,k,m,n) \in \mathbb{C}$ for all j,k,m,n,l with $\sum_j \sum_k \sum_{l=1}^{M_0} |a_0^{(l)}(j,k,m,n)|^2 <$

∞ for each m and n. We now show that $a_0^{(l)}$ depends on $m - j$. To see this, note that $U_1 P_{N_0} = P_{N_0} U_1$. Now, since

$$U_1 P_{N_0} X_{m,n} = \sum_{j \leq m} \sum_{k \leq n} \sum_{l=1}^{M_0} a_0^{(l)}(j,k,m,n) \nu_{j+1,k}^{(l)}$$

$$= \sum_{j \leq m+1} \sum_{k \leq n} \sum_{l=1}^{M_0} a_0^{(l)}(j-1,k,m,n) \nu_{j,k}^{(l)},$$

and

$$P_{N_0} U_1 X_{m,n} = P_{N_0} X_{m+1,n} = \sum_{j \leq m+1} \sum_{k \leq n} \sum_{l=1}^{M_0} a_0^{(l)}(j,k,m+1,n) \nu_{j,k}^{(l)}.$$

It follows that

$$\sum_{j \leq m+1} \sum_{k \leq n} \sum_{l=1}^{M_0} a_0^{(l)}(j-1,k,m,n) \nu_{j,k}^{(l)} = \sum_{j \leq m+1} \sum_{k \leq n} \sum_{l=1}^{M_0} a_0^{(l)}(j,k,m+1,n) \nu_{j,k}^{(l)}.$$

From this equation, we see that $a_0^{(l)}$ depends on $m-j$. An analogous argument will show that $a_0^{(l)}$ also depends on $n - k$. Henceforth, we will write $a_0^{(l)}$ as $a_0^{(l)}(m-j, n-k)$. Therefore, we have

$$P_{N_0} X_{m,n} = \sum_{j \leq m} \sum_{k \leq n} \sum_{l=1}^{M_0} a_0^{(l)}(m-j, n-k) \nu_{j,k}^{(l)},$$

where $a_0^{(l)}(j,k) \in \mathbb{C}$ for all j, k, l with $\displaystyle\sum_j \sum_k \sum_{l=1}^{M_0} |a_0^{(l)}(j,k)|^2 < \infty.$ $\qquad\square$

For the next lemma, we will use the notation $\ell^2(M)$, where M is either a positive integer or positive infinity. As usual, $\ell^2(M)$ will denote the collection of all sequences of complex numbers of length M that are square summable. It is well known that $\ell^2(M)$ endowed with the typical inner product is a Hilbert space.

Lemma 2.8.2. $P_{N_1} X_{m,n} = \displaystyle\sum_{j \leq m} \sum_{l=1}^{M_1} (a_1(m-j)C_1^n)^{(l)} \, \eta_j^{(l)},$ *where* $a_1^{(l)}(j) \in \mathbb{C}$

for all j, l *with* $\displaystyle\sum_j \sum_{l=1}^{M_1} |a_1^{(l)}(j)|^2 < \infty$, C_1 *is a unitary matrix on* $\ell^2(M_1)$ *and*

$\eta_j^{(l)} = U_1^j \eta_0^{(l)}$, *where* $\left\{\eta_0^{(l)}\right\}_{1 \leq l \leq M_1}$ *is an orthonormal basis for* I_0^1.

Proof. Let $\left\{\eta_0^{(l)}\right\}_{1 \leq l \leq M_1}$ is an orthonormal basis for I_0^1. Then, $\left\{\eta_j^{(l)}\right\}_{1 \leq l \leq M_1} = \left\{U_1^j \eta_0^{(l)}\right\}_{1 \leq l \leq M_1}$ is an orthonormal basis for I_j^1. Therefore,

$$P_{N_1} X_{m,n} = \sum_{j \leq m} \sum_{l=1}^{M_1} a_1^{(l)}(j, m, n) \eta_j^{(l)},$$

where $a_1^{(l)}(j, m, n) \in \mathbb{C}$ for all j, m, n, l with $\displaystyle\sum_j \sum_{l=1}^{M_1} |a_1^{(l)}(j, m, n)|^2 < \infty$ for each m and n.

Just as in the proof of Lemma 2.8.1, we get that $a_1^{(l)}$ depends on $m - j$. Therefore, we have

$$P_{N_1} X_{m,n} = \sum_{j \leq m} \sum_{l=1}^{M_1} a_1^{(l)}(m - j, n) \eta_j^{(l)}.$$

Now, since $U_2 P_{N_1} = P_{N_1} U_2$, we see that $U_2 \eta_j^{(l)} = \sum_{k=1}^{M_1} c_{k,l} \eta_j^{(k)}$ and

$$
\begin{aligned}
P_{N_1} X_{m,1} &= U_2 P_{N_1} X_{m,0} = \sum_{j \leq m} \sum_{l=1}^{M_1} a_1^{(l)}(m - j, 0) U_2 \eta_j^{(l)} \\
&= \sum_{j \leq m} \sum_{k=1}^{M_1} \sum_{l=1}^{M_1} a_1^{(l)}(m - j, 0) c_{k,l} \eta_j^{(k)} \\
&= \sum_{j \leq m} \sum_{k=1}^{M_1} (a_1(m - j, 0) C_1)^{(k)} \eta_j^{(k)},
\end{aligned}
$$

where C_1 is a $M_1 \times M_1$ matrix with entries $C_1 = [c_{k,l}]_{k,l}$. A straightforward calculation shows that C_1 is a unitary matrix. Continuing in this manner, we get

$$P_{N_1} X_{m,n} = \sum_{j \leq m} \sum_{k=1}^{M_1} (a_1(m-j)C_1^n)^{(k)} \eta_j^{(k)},$$

where $a_1(m-j) = a_1(m-j,0)$ and $\sum_j \sum_{l=1}^{M_1} |a_1^{(l)}(j)|^2 < \infty$. □

The next lemma follows using an analogous argument from the lemma just proved. For this reason, we do not include a proof.

Lemma 2.8.3. $P_{N_2} X_{m,n} = \sum_{k \leq n} \sum_{l=1}^{M_2} (a_2(n-k)C_2^m)^{(l)} \zeta_k^{(l)}$, *where* $a_2^{(l)}(k) \in \mathbb{C}$

for all k,l *with* $\sum_k \sum_{l=1}^{M_2} |a_2^{(l)}(k)|^2 < \infty$, C_2 *is a unitary matrix on* $\ell^2(M_2)$ *and*

$\zeta_k^{(l)} = U_2^k \zeta_0^{(l)}$, *where* $\left\{ \zeta_0^{(l)} \right\}_{1 \leq l \leq M_2}$ *is an orthonormal basis for* I_0^2.

The next two theorems follow immediately from these lemmas and Theorem 2.8.1. Before we state these theorems, we introduce two new terms. For a weakly stationary random field $X_{m,n}$, $(m,n) \in \mathbb{Z}^2$, we say it is

- weakly regular if $L^1(X : -\infty) \cap L^2(X : -\infty) = \{0\}$, and

- strongly regular if $L^1(X : -\infty) \bigvee L^2(X : -\infty) = \{0\}$.

Note: A weakly stationary random field is strongly regular if and only if it is both horizontally and vertically regular.

Theorem 2.8.4. *Let* $X_{m,n}$, $(m,n) \in \mathbb{Z}^2$ *be a weakly regular, weakly stationary random field satisfying (2.9), the weak commuting condition. Then,*

$$X_{m,n} = \sum_{j \leq m} \sum_{k \leq n} \sum_{l=1}^{M_0} a_0^{(l)}(m-j,n-k)\nu_{j,k}^{(l)}$$

$$+ \sum_{j \leq m} \sum_{l=1}^{M_1} (a_1(m-j)C_1^n)^{(l)} \eta_j^{(l)} + \sum_{k \leq n} \sum_{l=1}^{M_2} (a_2(n-k)C_2^m)^{(l)} \zeta_k^{(l)}$$

where the notation is defined in Lemmas 2.8.1, 2.8.2, and 2.8.3.

Theorem 2.8.5. *Let* $X_{m,n}$, $(m,n) \in \mathbb{Z}^2$ *be a strongly regular, weakly stationary random field satisfying (2.9), the weak commuting condition. Then,*

$$X_{m,n} = \sum_{j \leq m} \sum_{k \leq n} \sum_{l=1}^{M_0} a_0^{(l)}(m-j,n-k)\nu_{j,k}^{(l)}$$

where the notation is defined in Lemma 2.8.1.

Theorem 2.8.6. *Let $X_{m,n}$, $(m,n) \in \mathbb{Z}^2$ be a strongly regular, weakly station-ary random field satisfying (2.8), the strong commuting condition. Then,*

$$X_{m,n} = \sum_{(j,k) \in \mathbb{N}^2} a_{j,k}\, \xi_{m-j,n-k},$$

where $(a_{m,n})_{(m,n) \in \mathbb{N}^2}$ is a sequence of complex numbers with the property that
$$\sum_{(m,n) \in \mathbb{N}^2} |a_{m,n}|^2 < \infty \text{ and } \xi_{m,n},\ (m,n) \in \mathbb{Z}^2 \text{ is a white noise random field}$$
in $L(X)$. That is, $X_{m,n}$, $(m,n) \in \mathbb{Z}^2$ has a quarter-plane moving average representation.

Proof. It follows from Theorem 2.8.5 that all we need to prove is that under the additional condition (2.8), we get that the $\dim(W_{0,0}) = 1$. We can see this quickly by observing that under condition (2.8),

$$W_{0,0} \subseteq L(X:0,0) \subseteq \overline{span}\left\{L^1(X:-1) \vee L^2(X:-1) \vee \{X_{0,0}\}\right\}.$$

However, $L^1(X:-1) \perp W_{0,0}$ and $L^2(X:-1) \perp W_{0,0}$. Hence, $W_{0,0} = \{c\,X_{0,0} : c \in \mathbb{C}\}$ and so, $\dim(W_{0,0}) = 1$. $\qquad\square$

2.9   Quarter-plane Moving Average Representations

Let $(\Omega, \mathcal{F}, P)$ be a probability space. Let $X_{m,n}$, $(m,n) \in \mathbb{Z}^2$ be a random field. Suppose further that $X_{m,n}$, $(m,n) \in \mathbb{Z}^2$ is a quarter-plane moving average ran-dom field. That is, there exists a sequence of complex numbers $(a_{m,n})_{(m,n) \in \mathbb{N}^2}$ with the property that $\sum_{(m,n) \in \mathbb{N}^2} |a_{m,n}|^2 < \infty$ and a white noise random field $\xi_{m,n}$, $(m,n) \in \mathbb{Z}^2$ in $L^2(\Omega, \mathcal{F}, P)$ such that

$$X_{m,n} = \sum_{(j,k) \in \mathbb{N}^2} a_{j,k}\, \xi_{m-j,n-k},$$

in $L^2(\Omega, \mathcal{F}, P)$, for all $(m,n) \in \mathbb{Z}^2$. A straightforward calculation shows that $X_{m,n}$, $(m,n) \in \mathbb{Z}^2$ is a weakly stationary random field.

We begin by examining the spectral measure for this random field. First recall that white noise is a stationary random field and as such has a spectral representation.

$$\xi_{m,n} = \int_{[-\pi,\pi)} \int_{[-\pi,\pi)} e^{im\lambda + in\theta}\, d\mathcal{Z}_{\xi}(e^{i\lambda}, e^{i\theta}),$$

where $\mathcal{Z}_{\xi}$ is the orthogonally scattered set function associated with the random field $\xi_{m,n}$, $(m,n) \in \mathbb{Z}^2$. It is straightforward to verify that F_ξ, the spectral measure for $\xi_{m,n}$, $(m,n) \in \mathbb{Z}^2$, is normalized Lebesgue measure on $\mathbb{T}^2$. That is, $dF_\xi(e^{i\lambda}, e^{i\theta}) = d\sigma^2(e^{i\lambda}, e^{i\theta})$. We now calculate the covariance function for $X_{m,n}$, $(m,n) \in \mathbb{Z}^2$.

$$r_X(m,n) = (X_{m,n}, X_{0,0})_{L^2(\Omega,\mathcal{F},P)}$$

$$= \left(\sum_{(j,k)\in\mathbb{N}^2} a_{j,k}\, \xi_{m-j,n-k}, \sum_{(j,k)\in\mathbb{N}^2} a_{j,k}\, \xi_{-j,-k} \right)_{L^2(\Omega,\mathcal{F},P)}$$

$$= \int_{[-\pi,\pi)} \int_{[-\pi,\pi)} \sum_{(j,k)\in\mathbb{N}^2} a_{j,k}e^{i(m-j)\lambda + i(n-k)\theta} \overline{\sum_{(j,k)\in\mathbb{N}^2} a_{j,k}e^{-ij\lambda - ik\theta}}\, dF_\xi(e^{i\lambda}, e^{i\theta})$$

$$= \int_{[-\pi,\pi)} \int_{[-\pi,\pi)} \left| \sum_{(j,k)\in\mathbb{N}^2} a_{j,k}e^{-ij\lambda - ik\theta} \right|^2 e^{im\lambda + in\theta}\, d\sigma^2(e^{i\lambda}, e^{i\theta}).$$

Therefore, we see that the spectral measure of $X_{m,n}$, $(m,n) \in \mathbb{Z}^2$, is absolutely continuous with respect to normalized Lebesgue measure on $\mathbb{T}^2$, with

density $f_X(e^{i\lambda}, e^{i\theta}) = \left| \sum_{(j,k)\in\mathbb{N}^2} a_{j,k}\, e^{-ij\lambda - ik\theta} \right|^2 = \left| \sum_{(j,k)\in\mathbb{N}^2} \overline{a_{j,k}}\, e^{ij\lambda + ik\theta} \right|^2$. We now record this observation as a theorem.

Theorem 2.9.1. *Suppose that $X_{m,n}$, $(m,n) \in \mathbb{Z}^2$ is a quarter-plane moving average random field; that is, there exists a sequence of complex numbers $(a_{m,n})_{(m,n)\in\mathbb{N}^2}$ with the property that $\sum_{(m,n)\in\mathbb{N}^2} |a_{m,n}|^2 < \infty$ and a white noise random field $\xi_{m,n}$, $(m,n) \in \mathbb{Z}^2$ in $L^2(\Omega, \mathcal{F}, P)$ such that*

$$X_{m,n} = \sum_{(j,k)\in\mathbb{N}^2} a_{j,k}\, \xi_{m-j,n-k},$$

in $L^2(\Omega, \mathcal{F}, P)$, for all $(m,n) \in \mathbb{Z}^2$. Then, $X_{m,n}$, $(m,n) \in \mathbb{Z}^2$ is a weakly stationary random field that has a spectral measure that is absolutely continuous with respect to normalized Lebesgue measure with density $f_X(e^{i\lambda}, e^{i\theta}) = |\varphi(e^{i\lambda}, e^{i\theta})|^2$, where $\varphi \in H^2(\mathbb{T}^2)$ and $\hat{\varphi}(j,k) = \overline{a_{j,k}}$ for all $(j,k) \in \mathbb{N}^2$.

In the previous theorem, $H^2(\mathbb{T}^2) = \overline{span}\{e^{ik\lambda + il\theta} : (k,l) \in \mathbb{N}^2\}$ in $L^2(\mathbb{T}^2, \sigma^2)$.

Now, let us examine the situation from the other direction. Suppose that $X_{m,n}$, $(m,n) \in \mathbb{Z}^2$ is a weakly stationary random field with spectral measure $dF_X(e^{i\lambda}, e^{i\theta}) = |\varphi(e^{i\lambda}, e^{i\theta})|^2\, d\sigma^2(e^{i\lambda}, e^{i\theta})$, where $\varphi \in H^2(\mathbb{T}^2)$. As such, φ is equal to its Fourier series in $L^2(\mathbb{T}^2, \sigma^2)$. That is,

$$\varphi(e^{i\lambda}, e^{i\theta}) = \sum_{(j,k)\in\mathbb{N}^2} \hat{\varphi}(j,k)e^{ij\lambda + ik\theta} \quad \text{in } L^2(\mathbb{T}^2, \sigma^2).$$

We now define the linear transformation $W : L(X) \to L^2(\mathbb{T}^2, \sigma^2)$ by $W(X_{m,n}) = e^{im\lambda + in\theta}\overline{\varphi}(e^{i\lambda}, e^{i\theta})$. W is an isometry and as such, we may identify $L(X)$ and $W(L(X))$, as well as, $X_{m,n}$ and $e^{im\lambda + in\theta}\overline{\varphi}(e^{i\lambda}, e^{i\theta})$. $e^{im\lambda + in\theta}\overline{\varphi}(e^{i\lambda}, e^{i\theta})$ has a moving average representation since

$$e^{im\lambda + in\theta}\overline{\varphi}(e^{i\lambda}, e^{i\theta}) = \sum_{(j,k)\in\mathbb{N}^2} \overline{\hat{\varphi}}(j,k)e^{i(m-j)\lambda + i(n-k)\theta} \text{ in } L^2(\mathbb{T}^2, \sigma^2).$$

So, we may conclude the $X_{m,n}$, $(m,n) \in \mathbb{Z}^2$ has a moving average representation of the form

$$X_{m,n} = \sum_{(j,k)\in\mathbb{N}^2} \overline{\hat{\varphi}(j,k)}\xi_{m-j,n-k},$$

for some white noise random field $(\xi_{m,n})_{(m,n)\in\mathbb{Z}^2}$ contained in some Hilbert space that contains $L(X)$ as a subspace. A natural question at this point might be: when is this white noise sequence contained in $L(X)$? Let us examine that question. Let

$$L(\xi) = \overline{\text{span}}\left\{\xi_{m,n} : (m,n) \in \mathbb{Z}^2\right\}.$$

Now, define the linear transformation $W^{\#} : L(\xi) \to L^2(\mathbb{T}^2, \sigma^2)$ by $W^{\#}(\xi_{m,n}) = e^{im\lambda + in\theta}$. $W^{\#}$ is clearly an isomorphism. It follows from the moving average representation of $X_{m,n}$, $(m,n) \in \mathbb{Z}^2$, that $L(X) \subseteq L(\xi)$. Again from this moving average representation, we see that

$$W^{\#}(X_{m,n}) = e^{im\lambda + in\theta}\overline{\varphi}(e^{i\lambda}, e^{i\theta}).$$

It follows then that

$$W^{\#}(L(X)) = \overline{\text{span}}\left\{e^{im\lambda + in\theta}\overline{\varphi}(e^{i\lambda}, e^{i\theta}) : (m,n) \in \mathbb{Z}^2\right\},$$

which is a doubly invariant subspace of $L^2(\mathbb{T}^2)$. By Theorem 3.6.1, all doubly invariant subspaces of $L^2(\mathbb{T}^2, \sigma^2)$ are of the form $\mathbf{1}_E L^2(\mathbb{T}^2, \sigma^2)$, where E is a Lebesgue measurable subset of $\mathbb{T}^2$. In our particular case, $E = \left\{(e^{i\lambda}, e^{i\theta}) \in \mathbb{T}^2 : \varphi(e^{i\lambda}, e^{i\theta}) \neq 0\right\}$. From this observation, we see that if E^c, the complement of E, has Lebesgue measure zero, then $W^{\#}(L(X)) = L^2(\mathbb{T}^2, \sigma^2)$. This together with the fact that $W^{\#}$ is an isomorphism gives us that $L(X) = L(\xi)$ and therefore, $\xi_{m,n} \in L(X)$ for all $(m,n) \in \mathbb{Z}^2$. If, on the other hand, the Lebesgue measure of E^c is positive, then $L(X) \subsetneq L(\xi)$. So, there exists a $(j,k) \in \mathbb{Z}^2$ such that $\xi_{j,k} \notin L(X)$. However, since $H^2(\mathbb{T}^2)$ functions do not vanish on a set of positive Lebesgue measure, it follows that $\sigma^2(E^c) = 0$.

We will now record these observations as a theorem.

Theorem 2.9.2. *Suppose that $X_{m,n}$, $(m,n) \in \mathbb{Z}^2$ is a weakly stationary random field with spectral measure $dF_X(e^{i\lambda}, e^{i\theta}) = |\varphi(e^{i\lambda}, e^{i\theta})|^2 d\sigma^2(e^{i\lambda}, e^{i\theta})$, where $\varphi \in H^2(\mathbb{T}^2)$. Then, $X_{m,n}$, $(m,n) \in \mathbb{Z}^2$ may be represented as a quarter-plane moving average random field; that is,*

$$X_{m,n} = \sum_{(j,k)\in\mathbb{N}^2} \overline{\hat{\varphi}(j,k)}\xi_{m-j,n-k},$$

in $L^2(\Omega, \mathcal{F}, P)$, with our white noise random field $\xi_{m,n}$, $(m,n) \in \mathbb{Z}^2$ being contained in $L(X)$.

We now examine proper quarter-plane moving average random fields. First, we define

$$L(X : (m,n)) = \overline{\mathrm{span}}\{X_{j,k} : j \leq m, k \leq n\}.$$

Suppose that $X_{m,n}$, $(m,n) \in \mathbb{Z}^2$ is a quarter-plane moving average random field. Therefore, there exists a sequence of complex numbers $(a_{m,n})_{(m,n)\in\mathbb{N}^2}$ with the property that $\sum\limits_{(m,n)\in\mathbb{N}^2} |a_{m,n}|^2 < \infty$ and a white noise random field $\xi_{m,n}$, $(m,n) \in \mathbb{Z}^2$ in $L^2(\Omega, \mathcal{F}, P)$ such that

$$X_{m,n} = \sum_{(j,k)\in\mathbb{N}^2} a_{j,k}\, \xi_{m-j,n-k},$$

in $L^2(\Omega, \mathcal{F}, P)$, for all $(m,n) \in \mathbb{Z}^2$. Such a moving average is called **proper** if

$$L(X : (m,n)) = L(\xi : (m,n)), \text{ for all } (m,n) \in \mathbb{Z}^2.$$

From the moving average representation, we see that

$$W^{\#}(X_{m,n}) = e^{im\lambda+in\theta} \sum_{(j,k)\in\mathbb{N}^2} a_{j,k} e^{-ij\lambda-ik\theta},$$

where $W^{\#}$ is defined above. It follows from this calculation that

$$W^{\#}(L(X : (m,n))) = \overline{\mathrm{span}}\left\{e^{ij\lambda+ik\theta}f(e^{i\lambda}, e^{i\theta}) : j \leq m, k \leq n\right\} = e^{im\lambda+in\theta}[f]_{\mathbb{Z}^2_-},$$

where $f(e^{i\lambda}, e^{i\theta}) = \sum\limits_{(m,n)\in\mathbb{N}^2} a_{m,n} e^{-im\lambda-in\theta}$ and

$[f]_{\mathbb{Z}^2_-} = \overline{\mathrm{span}}\left\{e^{ij\lambda+ik\theta}f(e^{i\lambda}, e^{i\theta}) : j \leq 0, k \leq 0\right\}$. Similarly,

$$W^{\#}(L(\xi : (m,n))) = \overline{\mathrm{span}}\left\{e^{ij\lambda+ik\theta} : j \leq m, k \leq n\right\} = e^{im\lambda+in\theta}H^2_*(\mathbb{T}^2),$$

where $H^2_*(\mathbb{T}^2) = \overline{\mathrm{span}}\left\{e^{ij\lambda+ik\theta} : j \leq 0, k \leq 0\right\}$, the complex conjugate of $H^2(\mathbb{T}^2)$. It is straightforward to verify that $[f]_{\mathbb{Z}^2_-} = H^2_*(\mathbb{T}^2)$ is equivalent to $[\overline{f}]_{\mathbb{N}^2} = H^2(\mathbb{T}^2)$, where

$$[\overline{f}]_{\mathbb{N}^2} = \overline{\mathrm{span}}\left\{e^{ij\lambda+ik\theta}\overline{f}(e^{i\lambda}, e^{i\theta}) : j \geq 0, k \geq 0\right\}.$$

Notice that $\overline{f}$ is in $H^2(\mathbb{T}^2)$. A function g in $H^2(\mathbb{T}^2)$ is called **strongly outer** if $[g]_{\mathbb{N}^2} = H^2(\mathbb{T}^2)$. We summarize our findings in the following theorem.

Theorem 2.9.3. *Suppose that $X_{m,n}$, $(m,n) \in \mathbb{Z}^2$ is a quarter-plane moving average random field. Therefore, there exists a sequence of complex numbers*

$(a_{m,n})_{(m,n)\in\mathbb{N}^2}$ *with the property that* $\sum\limits_{(m,n)\in\mathbb{N}^2} |a_{m,n}|^2 < \infty$ *and a white noise random field* $\xi_{m,n}$, $(m,n) \in \mathbb{Z}^2$ *in* $L^2(\Omega, \mathcal{F}, P)$ *such that*

$$X_{m,n} = \sum_{(j,k)\in\mathbb{N}^2} a_{j,k}\, \xi_{m-j,n-k},$$

in $L^2(\Omega, \mathcal{F}, P)$, *for all* $(m,n) \in \mathbb{Z}^2$. *This moving average representation is proper if and only if* φ *is strongly outer, where* $\varphi(e^{i\lambda}, e^{i\theta}) = \sum\limits_{(m,n)\in\mathbb{N}^2} \bar{a}_{m,n} e^{im\lambda + in\theta}$.

We now examine quarter-plane moving average random fields in the context of regularity.

Theorem 2.9.4. *If* $X_{m,n}$, $(m,n) \in \mathbb{Z}^2$ *is a quarter-plane moving average random field, then* $X_{m,n}$, $(m,n) \in \mathbb{Z}^2$ *is both horizontally and vertically regular.*

Proof. In Subsection 2.4.3, we showed that every white noise random field is both horizontally and vertically regular. Therefore, since

$$X_{m,n} = \sum_{(j,k)\in\mathbb{N}^2} a_{j,k}\, \xi_{m-j,n-k},$$

where $(a_{m,n})_{(m,n)\in\mathbb{N}^2}$ is a sequence of complex numbers with $\sum\limits_{(m,n)\in\mathbb{N}^2} |a_{m,n}|^2 < \infty$ and $\xi_{m,n}$, $(m,n) \in \mathbb{Z}^2$ is a white noise random field. It then follows that $L^j(X : m) \subseteq L^j(\xi : m)$, for all $m \in \mathbb{Z}$ and $j = 1, 2$. Hence, $X_{m,n}$, $(m,n) \in \mathbb{Z}^2$ is both horizontally and vertically regular. $\qquad\square$

Theorem 2.9.5. *If* $X_{m,n}$, $(m,n) \in \mathbb{Z}^2$ *is a proper quarter-plane moving average random field, then the strong commuting condition holds on* $L(X)$.

Note: The strong commuting condition is given in Equation (2.8).

Proof. Suppose $X_{m,n}$, $(m,n) \in \mathbb{Z}^2$ is a proper quarter-plane moving average random field. Then, by definition,

$$X_{m,n} = \sum_{(j,k)\in\mathbb{N}^2} a_{j,k}\, \xi_{m-j,n-k},$$

where $(a_{m,n})_{(m,n)\in\mathbb{N}^2}$ is a sequence of complex numbers with $\sum\limits_{(m,n)\in\mathbb{N}^2} |a_{m,n}|^2 < \infty$ and $\xi_{m,n}$, $(m,n) \in \mathbb{Z}^2$ is a white noise random field, with $L(X) = L(\xi)$, $L^j(X : m) = L^j(\xi : m)$, for all $m \in \mathbb{Z}$ and $j = 1, 2$ and $L(X : (m,n)) = L(\xi : (m,n))$ for all $(m,n) \in \mathbb{Z}^2$. It follows from these observations that we need only show that the strong commuting condition holds on $L(\xi)$. This is straightforward to see since

$$L^1(\xi : m) = L(\xi : (m,n)) \oplus \overline{span}\{\xi_{k,l} : k \leq m, l > n\}$$

and

$$L^2(\xi : n) = L(\xi : (m,n)) \oplus \overline{span}\{\xi_{k,l} : k > m, l \leq n\}.$$

If we let $M_1 = \overline{span}\{\xi_{k,l} : k \leq m, l > n\}$ and $M_2 = \overline{span}\{\xi_{k,l} : k > m, l \leq n\}$, we may write

$$P_{L^1(\xi:m)} = P_{L(\xi:(m,n))} + P_{M_1}$$

and

$$P_{L^2(\xi:n)} = P_{L(\xi:(m,n))} + P_{M_2}.$$

Finally, observing that M_1 is orthogonal to M_2, we conclude that $P_{L^1(\xi:m)} P_{L^2(\xi:n)} = P_{L(\xi:(m,n))}$. That is, the strong commuting condition holds on $L(\xi)$. $\qquad\square$

Theorem 2.9.6. *If $X_{m,n}$, $(m,n) \in \mathbb{Z}^2$ is a quarter-plane moving average random field and the strong commuting condition holds on $L(X)$, then $X_{m,n}$, $(m,n) \in \mathbb{Z}^2$ can be written as a proper quarter-plane moving average random field.*

Proof. Since $X_{m,n}$, $(m,n) \in \mathbb{Z}^2$ is a quarter-plane moving average random field, we know that

$$X_{m,n} = \sum_{(j,k)\in\mathbb{N}^2} a_{j,k}\, \xi_{m-j,n-k},$$

where $(a_{m,n})_{(m,n)\in\mathbb{N}^2}$ is a sequence of complex numbers with $\displaystyle\sum_{(m,n)\in\mathbb{N}^2} |a_{m,n}|^2 <$ ∞ and $\xi_{m,n}$, $(m,n) \in \mathbb{Z}^2$ is a white noise random field. By Theorem 2.11.1, we know that $X_{m,n}$, $(m,n) \in \mathbb{Z}^2$ has a spectral measure that is absolutely continuous with respect to normalized Lebesgue measure with density $f_X(e^{i\lambda}, e^{i\theta}) = |\varphi(e^{i\lambda}, e^{i\theta})|^2$, where $\varphi \in H^2(\mathbb{T}^2)$ and $\hat{\varphi}(j,k) = \overline{a_{j,k}}$ for all $(j,k) \in \mathbb{N}^2$. It was shown in [25] that the strong commuting condition implies that $\varphi = fg$, where f is inner and g is strongly outer. Since inner functions have modulus one $[\sigma^2]$-a.e. on $\mathbb{T}^2$, it follows that $f_X(e^{i\lambda}, e^{i\theta}) = |g(e^{i\lambda}, e^{i\theta})|^2$. Therefore, by Theorem 2.9.2, it follows that

$$X_{m,n} = \sum_{(j,k)\in\mathbb{N}^2} \overline{\hat{g}(j,k)}\, \eta_{m-j,n-k},$$

where $\eta_{m,n}$, $(m,n) \in \mathbb{Z}^2$ is a white noise random field. It then follows from Theorem 2.9.3 that this representation is proper. $\qquad\square$

Theorem 2.9.6 and Theorem 2.8.6 together give us the following theorem.

Theorem 2.9.7. *Let $X_{m,n}$, $(m,n) \in \mathbb{Z}^2$ be a strongly regular, weakly stationary random field satisfying (2.8), the strong commuting condition. Then, $X_{m,n}$, $(m,n) \in \mathbb{Z}^2$ has a proper quarter-plane moving average representation.*

We now organize some of the results from this section into the following theorem.

Theorem 2.9.8. *Let $X_{m,n}$, $(m,n) \in \mathbb{Z}^2$ be a weakly stationary random field with spectral measure F_X. Then, the following are equivalent.*

    1. $X_{m,n}$, $(m,n) \in \mathbb{Z}^2$ is strongly regular and satisfies (2.8), the strong commuting condition.

    2. F_X is absolutely continuous with respect to σ^2 with density $f_X(e^{i\lambda}, e^{i\theta}) = \left| \varphi(e^{i\lambda}, e^{i\theta}) \right|^2$, where φ is a strongly outer function in $H^2(\mathbb{T}^2)$.

    3. $X_{m,n}$, $(m,n) \in \mathbb{Z}^2$ has a proper quarter-plane moving average representation.

Note that these equivalent conditions imply that F_X is absolutely continuous with respect to σ^2 with density $f_X \geq 0$ $[\sigma^2]$-a.e. satisfying

$$\int_{[-\pi,\pi)} \int_{[-\pi,\pi)} \log(f_X(e^{i\lambda}, e^{i\theta}))\, d\sigma^2(e^{i\lambda}, e^{i\theta}) > -\infty.$$

Although this condition does imply that F_X is absolutely continuous with respect to σ^2 with density $f_X(e^{i\lambda}, e^{i\theta}) = \left| \varphi(e^{i\lambda}, e^{i\theta}) \right|^2$, where φ is in $H^2(\mathbb{T}^2)$. It does not imply that φ is strongly outer.

2.10   Helson-Lowdenslager Theory

In [10], Helson and Lowdenslager studied perdition theory and Fourier series in several variables. Their work centered around analyticity in a group-theoretic context. In this section, we present the main ideas of their work centered around Fourier series in several variables. In a later section, we will present the connections to weakly stationary random fields.

Let S be a subset of $\mathbb{Z}^2$ with the following properties:

(a) $S \cup (-S) = \mathbb{Z}^2$,

(b) $S \cap (-S) = \{(0,0)\}$, and

(c) if (m,n) and (m',n') are in S, then $(m+m', n+n')$ is in S.

Any subset with these properties is called a **semigroup** of $\mathbb{Z}^2$.

Before we state the main result of this section, we make some observations that will be used in the proof.

Let $H^2_{S_0}(\mathbb{T}^2, \mu)$ denote the closure in $L^2(\mathbb{T}^2, \mu)$, the Hilbert space of square-integrable function on $\mathbb{T}^2$ with respect to the finite nonnegative measure μ, of all trigonometric polynomials of the form

$$P(e^{i\lambda}, e^{i\theta}) = \sum_{(m,n)\in S_0} a_{m,n} e^{-i(m\lambda+n\theta)},$$

where $S_0 = S \setminus \{(0,0)\}$. Let $1 + H^2_{S_0}(\mathbb{T}^2, \mu)$ denote the set of all elements of the form $1 + f$, where f is a member of $H^2_{S_0}(\mathbb{T}^2, \mu)$. $1 + H^2_{S_0}(\mathbb{T}^2, \mu)$ forms a nonempty, closed, convex subset of $L^2(\mathbb{T}^2, \mu)$. As such, it contains a unique element of smallest norm. If this element is zero, then we say that prediction is perfect. Otherwise, we get a nonzero element of minimal norm of the form $1 + H$, where H is in $H^2_{S_0}(\mathbb{T}^2, \mu)$.

For any complex number c and (m, n) in S_0,

$$1 + H(e^{i\lambda}, e^{i\theta}) + c e^{-i(m\lambda+n\theta)}$$

belongs to $1 + H^2_{S_0}(\mathbb{T}^2, \mu)$. Since $1 + H$ is the unique element of minimal norm in $1 + H^2_{S_0}(\mathbb{T}^2, \mu)$, we know that

$$\int_{[-\pi,\pi)} \int_{[-\pi,\pi)} \left| 1 + H(e^{i\lambda}, e^{i\theta}) + c e^{-i(m\lambda+n\theta)} \right|^2 d\mu(e^{i\lambda}, e^{i\theta})$$

has a unique minimum at $c = 0$. A straightforward calculation then shows that for every (m, n) in S_0,

$$\int_{[-\pi,\pi)} \int_{[-\pi,\pi)} \left(1 + H(e^{i\lambda}, e^{i\theta})\right) e^{i(m\lambda+n\theta)} \, d\mu(e^{i\lambda}, e^{i\theta}) = 0. \qquad (2.15)$$

Since S is closed under addition, we get a second orthogonality relation. For each complex number c and each (m, n) in S_0, the function

$$\left(1 + H(e^{i\lambda}, e^{i\theta})\right) \left(1 + c e^{-i(m\lambda+n\theta)}\right)$$

belongs to $1 + H^2_{S_0}(\mathbb{T}^2, \mu)$, and its norm is minimized at $c = 0$. Therefore,

$$\int_{[-\pi,\pi)} \int_{[-\pi,\pi)} \left| 1 + H(e^{i\lambda}, e^{i\theta}) \right|^2 e^{i(m\lambda+n\theta)} \, d\mu(e^{i\lambda}, e^{i\theta}) = 0, \qquad (2.16)$$

for all (m, n) in S_0. By taking the complex conjugate of (2.16), we get that the same equation holds for all (m, n) in $-S_0$. That is, the Fourier-Stieltjes coefficients of the measure $\left|1 + H(e^{i\lambda}, e^{i\theta})\right|^2 d\mu(e^{i\lambda}, e^{i\theta})$ all vanish except the central one. Therefore, this measure is a multiple of Lebesgue measure. It follows that $1 + H$ must vanish almost everywhere with respect to the singular component of $d\mu$, and (2.15) can be written as,

$$\int_{[-\pi,\pi)} \int_{[-\pi,\pi)} \left(1 + H(e^{i\lambda}, e^{i\theta})\right) e^{i(m\lambda + n\theta)} d\mu_a(e^{i\lambda}, e^{i\theta}) = 0, \qquad (m,n) \in S_0,$$

(2.17)

where μ_a is the absolutely continuous part of μ with respect to Lebesgue measure.

Next, we observe that (2.15) characterizes the minimal element in $1 + H_{S_0}^2(\mathbb{T}^2, \mu)$. To see this, suppose that (2.15) holds, but $1 + G$ is the minimal element. Then,

$$\int_{[-\pi,\pi)} \int_{[-\pi,\pi)} \left|1 + H(e^{i\lambda}, e^{i\theta}) + c\left(G(e^{i\lambda}, e^{i\theta}) - H(e^{i\lambda}, e^{i\theta})\right)\right|^2 d\mu(e^{i\lambda}, e^{i\theta})$$

$$= \int_{[-\pi,\pi)} \int_{[-\pi,\pi)} \left|1 + H(e^{i\lambda}, e^{i\theta})\right|^2 d\mu(e^{i\lambda}, e^{i\theta})$$

$$+ |c|^2 \int_{[-\pi,\pi)} \int_{[-\pi,\pi)} \left|G(e^{i\lambda}, e^{i\theta}) - H(e^{i\lambda}, e^{i\theta})\right|^2 d\mu(e^{i\lambda}, e^{i\theta}),$$

for every complex number c. The right-hand form of the expression is clearly smallest when $c = 0$, and the left-hand form of the expression is certainly smallest when $c = 1$ since $1 + G$ is the minimal element. From these observations and the fact that the minimal element is unique, we conclude that $G = H$. We now state the main result of this section.

Theorem 2.10.1 (A Generalization of the Szegö-Krein-Kolmogorov Theorem). *Let S be a semigroup of $\mathbb{Z}^2$ and let μ be a finite nonnegative measure on $\mathbb{T}^2$. Let μ have Lebesgue decomposition*

$$d\mu(e^{i\lambda}, e^{i\theta}) = w(e^{i\lambda}, e^{i\theta}) \, d\sigma^2(e^{i\lambda}, e^{i\theta}) + d\mu_s(e^{i\lambda}, e^{i\theta}),$$

where w is nonnegative and in $L^1(\mathbb{T}^2, \sigma^2)$, the collection of all Lebesgue integrable functions on $\mathbb{T}^2$, and μ_s is singular with respect to σ^2. Then,

$$\exp\left\{\int_{[-\pi,\pi)} \int_{[-\pi,\pi)} \log w(e^{i\lambda}, e^{i\theta}) \, d\sigma^2(e^{i\lambda}, e^{i\theta})\right\}$$

$$= \inf_P \int_{[-\pi,\pi)} \int_{[-\pi,\pi)} \left|1 + P(e^{i\lambda}, e^{i\theta})\right|^2 d\mu(e^{i\lambda}, e^{i\theta}), \qquad (2.18)$$

where P ranges over finite sums of the form

$$P(e^{i\lambda}, e^{i\theta}) = \sum_{(m,n) \in S_0} a_{m,n} e^{-i(m\lambda + n\theta)},$$

where $S_0 = S \setminus \{(0,0)\}$. The left side of Equation (2.18) is interpreted as zero if

$$\int_{[-\pi,\pi)} \int_{[-\pi,\pi)} \log w(e^{i\lambda}, e^{i\theta}) \, d\sigma^2(e^{i\lambda}, e^{i\theta}) = -\infty.$$

Before we give a proof of this theorem, we make a few more observations and establish two lemmas. The proof of this theorem and the following lemmas are due to Helson and Lowdenslager, see [10].

If the infimum in (2.18) is positive, we have established the fact that it is equal to

$$\int_{[-\pi,\pi)} \int_{[-\pi,\pi)} \left| 1 + H(e^{i\lambda}, e^{i\theta}) \right|^2 \, d\mu(e^{i\lambda}, e^{i\theta}),$$

where $1 + H$ belongs to $1 + H^2_{S_0}(\mathbb{T}^2, \mu)$ and vanishes almost everywhere with respect to the singular component of $d\mu$. Hence, (2.17) holds. Moreover, $1 + H$ belongs to the convex set $1 + H^2_{S_0}(\mathbb{T}^2, w \, d\sigma^2)$. Note the closure is now with respect to $w \, d\sigma^2$ rather than μ. Furthermore, (2.17) implies that $1 + H$ is the minimal function relative to this measure and so

$$\inf_P \int_{[-\pi,\pi)} \int_{[-\pi,\pi)} \left| 1 + P(e^{i\lambda}, e^{i\theta}) \right|^2 w(e^{i\lambda}, e^{i\theta}) \, d\sigma^2(e^{i\lambda}, e^{i\theta})$$

$$= \int_{[-\pi,\pi)} \int_{[-\pi,\pi)} \left| 1 + H(e^{i\lambda}, e^{i\theta}) \right|^2 w(e^{i\lambda}, e^{i\theta}) \, d\sigma^2(e^{i\lambda}, e^{i\theta})$$

$$= \int_{[-\pi,\pi)} \int_{[-\pi,\pi)} \left| 1 + H(e^{i\lambda}, e^{i\theta}) \right|^2 \, d\mu(e^{i\lambda}, e^{i\theta}).$$

Therefore, it will suffice to prove that

$$\exp\left\{ \int_{[-\pi,\pi)} \int_{[-\pi,\pi)} \log w(e^{i\lambda}, e^{i\theta}) \, d\sigma^2(e^{i\lambda}, e^{i\theta}) \right\}$$

$$= \inf_P \int_{[-\pi,\pi)} \int_{[-\pi,\pi)} \left| 1 + P(e^{i\lambda}, e^{i\theta}) \right|^2 w(e^{i\lambda}, e^{i\theta}) \, d\sigma^2(e^{i\lambda}, e^{i\theta}). \qquad (2.19)$$

On the other hand, if the infimum in (2.18) is zero, it is enough to prove (2.19). In this case, the infimum in (2.19) will also vanish and having proved (2.19), (2.18) will follow. So, we shall prove (2.19) for an arbitrary nonnegative function w in $L^1(\mathbb{T}^2, \sigma^2)$.

Lemma 2.10.1. *If w is nonnegative and in $L^1(\mathbb{T}^2, \sigma^2)$, then*

$$\exp\left\{\int_{[-\pi,\pi)}\int_{[-\pi,\pi)}\log w(e^{i\lambda},e^{i\theta})\,d\sigma^2(e^{i\lambda},e^{i\theta})\right\}$$

$$= \inf_\psi \int_{[-\pi,\pi)}\int_{[-\pi,\pi)} e^{\psi(e^{i\lambda},e^{i\theta})}\, w(e^{i\lambda},e^{i\theta})\,d\sigma^2(e^{i\lambda},e^{i\theta}),$$

where ψ ranges over the real-valued functions in $L^1(\mathbb{T}^2,\sigma^2)$ such that

$$\hat{\psi}(0,0) := \int_{[-\pi,\pi)}\int_{[-\pi,\pi)} \psi(e^{i\lambda},e^{i\theta})\,d\sigma^2(e^{i\lambda},e^{i\theta}) = 0.$$

Proof. First, suppose that $\log w$ in $L^1(\mathbb{T}^2,\sigma^2)$. Then, we have

$$\exp\left\{\int_{[-\pi,\pi)}\int_{[-\pi,\pi)}\log w(e^{i\lambda},e^{i\theta})\,d\sigma^2(e^{i\lambda},e^{i\theta})\right\}$$

$$= \exp\left\{\int_{[-\pi,\pi)}\int_{[-\pi,\pi)}\log e^{\psi(e^{i\lambda},e^{i\theta})}\,d\sigma^2(e^{i\lambda},e^{i\theta})\right.$$

$$\left. + \int_{[-\pi,\pi)}\int_{[-\pi,\pi)}\log w(e^{i\lambda},e^{i\theta})\,d\sigma^2(e^{i\lambda},e^{i\theta})\right\}$$

$$= \exp\left\{\int_{[-\pi,\pi)}\int_{[-\pi,\pi)}\log\left(e^{\psi(e^{i\lambda},e^{i\theta})}w(e^{i\lambda},e^{i\theta})\right)\,d\sigma^2(e^{i\lambda},e^{i\theta})\right\}$$

$$\leq \int_{[-\pi,\pi)}\int_{[-\pi,\pi)} e^{\psi(e^{i\lambda},e^{i\theta})}w(e^{i\lambda},e^{i\theta})\,d\sigma^2(e^{i\lambda},e^{i\theta}).$$

The last inequality follows from Jensen's Inequality. Therefore,

$$\exp\left\{\int_{[-\pi,\pi)}\int_{[-\pi,\pi)}\log w(e^{i\lambda},e^{i\theta})\,d\sigma^2(e^{i\lambda},e^{i\theta})\right\}$$

$$\leq \inf_\psi \int_{[-\pi,\pi)}\int_{[-\pi,\pi)} e^{\psi(e^{i\lambda},e^{i\theta})}w(e^{i\lambda},e^{i\theta})\,d\sigma^2(e^{i\lambda},e^{i\theta}),$$

where the infimum is taken over all real-valued functions ψ in $L^1(\mathbb{T}^2,\sigma^2)$ such that $\hat{\psi}(0,0) = 0$. To see that equality holds, let

$$\psi(e^{i\lambda},e^{i\theta}) = \int_{[-\pi,\pi)}\int_{[-\pi,\pi)}\log w(e^{i\lambda},e^{i\theta})\,d\sigma^2(e^{i\lambda},e^{i\theta}) - \log w(e^{i\lambda},e^{i\theta}).$$

Then ψ satisfies all of the conditions of the lemma and

$$\exp\left\{\int_{[-\pi,\pi)}\int_{[-\pi,\pi)}\log w(e^{i\lambda},e^{i\theta})\,d\sigma^2(e^{i\lambda},e^{i\theta})\right\}$$

$$= \int_{[-\pi,\pi)}\int_{[-\pi,\pi)} e^{\psi(e^{i\lambda},e^{i\theta})}w(e^{i\lambda},e^{i\theta})\,d\sigma^2(e^{i\lambda},e^{i\theta}).$$

Therefore, $\exp\left\{\displaystyle\int_{[-\pi,\pi)}\int_{[-\pi,\pi)}\log w(e^{i\lambda},e^{i\theta})\,d\sigma^2(e^{i\lambda},e^{i\theta})\right\}$

$$= \inf_{\psi}\int_{[-\pi,\pi)}\int_{[-\pi,\pi)}e^{\psi(e^{i\lambda},e^{i\theta})}w(e^{i\lambda},e^{i\theta})\,d\sigma^2(e^{i\lambda},e^{i\theta}).$$

Now, suppose that $\log w$ is not in $L^1(\mathbb{T}^2,\sigma^2)$. Then, $\log(w+\epsilon)$ is in $L^1(\mathbb{T}^2,\sigma^2)$, for every $\epsilon > 0$. Therefore, it follows from above that

$\exp\left\{\displaystyle\int_{[-\pi,\pi)}\int_{[-\pi,\pi)}\log(w(e^{i\lambda},e^{i\theta})+\epsilon)\,d\sigma^2(e^{i\lambda},e^{i\theta})\right\}$

$$= \inf_{\psi}\int_{[-\pi,\pi)}\int_{[-\pi,\pi)}e^{\psi(e^{i\lambda},e^{i\theta})}(w(e^{i\lambda},e^{i\theta})+\epsilon)\,d\sigma^2(e^{i\lambda},e^{i\theta})$$

$$\geq \inf_{\psi}\int_{[-\pi,\pi)}\int_{[-\pi,\pi)}e^{\psi(e^{i\lambda},e^{i\theta})}w(e^{i\lambda},e^{i\theta})\,d\sigma^2(e^{i\lambda},e^{i\theta}) \geq 0.$$

By the Monotone Limit Theorem, the left-hand side of the above equation converges to

$$\exp\left\{\int_{[-\pi,\pi)}\int_{[-\pi,\pi)}\log w(e^{i\lambda},e^{i\theta})\,d\sigma^2(e^{i\lambda},e^{i\theta})\right\} = 0,$$

as ϵ tends to zero. Therefore, in this case as well,

$\exp\left\{\displaystyle\int_{[-\pi,\pi)}\int_{[-\pi,\pi)}\log w(e^{i\lambda},e^{i\theta})\,d\sigma^2(e^{i\lambda},e^{i\theta})\right\}$

$$= \inf_{\psi}\int_{[-\pi,\pi)}\int_{[-\pi,\pi)}e^{\psi(e^{i\lambda},e^{i\theta})}w(e^{i\lambda},e^{i\theta})\,d\sigma^2(e^{i\lambda},e^{i\theta}).$$

$\square$

Lemma 2.10.2. *If w is nonnegative and in $L^1(\mathbb{T}^2,\sigma^2)$, then*

$\exp\left\{\displaystyle\int_{[-\pi,\pi)}\int_{[-\pi,\pi)}\log w(e^{i\lambda},e^{i\theta})\,d\sigma^2(e^{i\lambda},e^{i\theta})\right\}$

$$= \inf_{\psi}\int_{[-\pi,\pi)}\int_{[-\pi,\pi)}e^{\psi(e^{i\lambda},e^{i\theta})}\,w(e^{i\lambda},e^{i\theta})\,d\sigma^2(e^{i\lambda},e^{i\theta}),$$

where ψ ranges over the real-valued trigonometric polynomials such that

$$\hat{\psi}(0,0) := \int_{[-\pi,\pi)}\int_{[-\pi,\pi)}\psi(e^{i\lambda},e^{i\theta})\,d\sigma^2(e^{i\lambda},e^{i\theta}) = 0.$$

Proof. We will assume that $\log w$ in $L^1(\mathbb{T}^2,\sigma^2)$, since we may use a limiting argument, as in Lemma 2.10.1, for the general case. It follows from the proof of Lemma 2.10.1 that

$$\exp\left\{\int_{[-\pi,\pi)}\int_{[-\pi,\pi)}\log w(e^{i\lambda},e^{i\theta})\,d\sigma^2(e^{i\lambda},e^{i\theta})\right\}$$

$$\leq \inf_{\psi}\int_{[-\pi,\pi)}\int_{[-\pi,\pi)}e^{\psi(e^{i\lambda},e^{i\theta})}w(e^{i\lambda},e^{i\theta})\,d\sigma^2(e^{i\lambda},e^{i\theta}),$$

where the infimum is taken over all real-valued trigonometric polynomials ψ such that $\hat{\psi}(0,0)=0$. For the other direction, we assume that

$$\int_{[-\pi,\pi)}\int_{[-\pi,\pi)}\log w(e^{i\lambda},e^{i\theta})\,d\sigma^2(e^{i\lambda},e^{i\theta})=0.$$

If this were not the case, we can divide f by

$$k=\exp\left\{\int_{[-\pi,\pi)}\int_{[-\pi,\pi)}\log w(e^{i\lambda},e^{i\theta})\,d\sigma^2(e^{i\lambda},e^{i\theta})\right\}. \text{ Then,}$$

$$\int_{[-\pi,\pi)}\int_{[-\pi,\pi)}\log\frac{w(e^{i\lambda},e^{i\theta})}{k}\,d\sigma^2(e^{i\lambda},e^{i\theta})=0.$$

We could then work out the details of the proof with w/k in place of w. Since,

$$\int_{[-\pi,\pi)}\int_{[-\pi,\pi)}\log w(e^{i\lambda},e^{i\theta})\,d\sigma^2(e^{i\lambda},e^{i\theta})=0.$$

It remains to show that

$$\inf_{\psi}\int_{[-\pi,\pi)}\int_{[-\pi,\pi)}e^{\psi(e^{i\lambda},e^{i\theta})}w(e^{i\lambda},e^{i\theta})\,d\sigma^2(e^{i\lambda},e^{i\theta})\leq 1,$$

where the infimum is taken over all real-valued trigonometric polynomials ψ such that $\hat{\psi}(0,0)=0$. Since every bounded function ψ is boundedly the limit of Fejér means of its Fourier series, with each approximating function P a trigonometric polynomial that is real-valued if ψ is real-valued, and $\hat{P}(0,0)=0$ if $\hat{\psi}(0,0)=0$, we will show this final inequality for bounded real-valued functions ψ and then our desired result follows.

Let $\left(u_n(e^{i\lambda},e^{i\theta})\right)_{n=1}^{\infty}$ be a sequence of nonnegative bounded functions that increase pointwise to $\log^+ w(e^{i\lambda},e^{i\theta})$, and let $\left(v_n(e^{i\lambda},e^{i\theta})\right)_{n=1}^{\infty}$ be a sequence of nonnegative bounded functions that increase pointwise to $\log^- w(e^{i\lambda},e^{i\theta})$. Then, by the monotone limit theorem,

$$\lim_{n\to\infty}\int_{[-\pi,\pi)}\int_{[-\pi,\pi)}u_n(e^{i\lambda},e^{i\theta})\,d\sigma^2(e^{i\lambda},e^{i\theta})$$

$$=\int_{[-\pi,\pi)}\int_{[-\pi,\pi)}\log^+ w(e^{i\lambda},e^{i\theta})\,d\sigma^2(e^{i\lambda},e^{i\theta})$$

$$= \int_{[-\pi,\pi)} \int_{[-\pi,\pi)} \log^{-} w(e^{i\lambda}, e^{i\theta}) \, d\sigma^{2}(e^{i\lambda}, e^{i\theta})$$

$$= \lim_{n\to\infty} \int_{[-\pi,\pi)} \int_{[-\pi,\pi)} v_{n}(e^{i\lambda}, e^{i\theta}) \, d\sigma^{2}(e^{i\lambda}, e^{i\theta}).$$

It then follows that for each n, there exists an m such that

$$\int_{[-\pi,\pi)} \int_{[-\pi,\pi)} u_{n}(e^{i\lambda}, e^{i\theta}) \, d\sigma^{2}(e^{i\lambda}, e^{i\theta}) \le \int_{[-\pi,\pi)} \int_{[-\pi,\pi)} v_{m}(e^{i\lambda}, e^{i\theta}) \, d\sigma^{2}(e^{i\lambda}, e^{i\theta}).$$

We can multiply v_{m} by a positive constant less than or equal to one, and rename the function v_{n}, so that

$$\int_{[-\pi,\pi)} \int_{[-\pi,\pi)} u_{n}(e^{i\lambda}, e^{i\theta}) \, d\sigma^{2}(e^{i\lambda}, e^{i\theta}) = \int_{[-\pi,\pi)} \int_{[-\pi,\pi)} v_{n}(e^{i\lambda}, e^{i\theta}) \, d\sigma^{2}(e^{i\lambda}, e^{i\theta}).$$

This new sequence $\left(v_{n}(e^{i\lambda}, e^{i\theta})\right)_{n=1}^{\infty}$ still converges pointwise to $\log^{-} w(e^{i\lambda}, e^{i\theta})$, although the convergence may no longer be monotonic. It follows from the construction that

$$0 \le e^{(\log^{+} w - u_{n}) - (\log^{-} w - v_{n})} \le \max\{1, w\}.$$

Therefore, we can apply the Lebesgue dominated convergence theorem to get

$$\lim_{n\to\infty} \int_{[-\pi,\pi)} \int_{[-\pi,\pi)} e^{(\log^{+} w(e^{i\lambda}, e^{i\theta}) - u_{n}(e^{i\lambda}, e^{i\theta})) - (\log^{-} w(e^{i\lambda}, e^{i\theta}) - v_{n}(e^{i\lambda}, e^{i\theta}))}$$

$$\times \, d\sigma^{2}(e^{i\lambda}, e^{i\theta}) = 1.$$

Since

$$\int_{[-\pi,\pi)} \int_{[-\pi,\pi)} e^{(\log^{+} w(e^{i\lambda}, e^{i\theta}) - u_{n}(e^{i\lambda}, e^{i\theta})) - (\log^{-} w(e^{i\lambda}, e^{i\theta}) - v_{n}(e^{i\lambda}, e^{i\theta}))} \, d\sigma^{2}(e^{i\lambda}, e^{i\theta})$$

$$= \int_{[-\pi,\pi)} \int_{[-\pi,\pi)} e^{v_{n}(e^{i\lambda}, e^{i\theta}) - u_{n}(e^{i\lambda}, e^{i\theta})} w(e^{i\lambda}, e^{i\theta}) \, d\sigma^{2}(e^{i\lambda}, e^{i\theta}),$$

$v_{n} - u_{n}$ is a real-valued bounded function with $\widehat{(v_{n} - u_{n})}(0,0) = 0$, it follows that

$$\inf_{\psi} \int_{[-\pi,\pi)} \int_{[-\pi,\pi)} e^{\psi(e^{i\lambda}, e^{i\theta})} w(e^{i\lambda}, e^{i\theta}) \, d\sigma^{2}(e^{i\lambda}, e^{i\theta}) \le 1,$$

where the infimum is taken over all real-valued bounded functions ψ such that $\hat{\psi}(0,0) = 0$. $\qquad\square$

We are now ready to prove Theorem 2.10.1.

Proof. First, real-valued trigonometric polynomials satisfying the conditions of Lemma 2.10.2 can be represented in the form $P(e^{i\lambda}, e^{i\theta}) + \overline{P(e^{i\lambda}, e^{i\theta})}$, where $P(e^{i\lambda}, e^{i\theta}) = \sum_{(m,n) \in S_0} a_{m,n} e^{-i(m\lambda + n\theta)}$. Therefore, we may write

$$e^{\psi(e^{i\lambda}, e^{i\theta})} = e^{P(e^{i\lambda}, e^{i\theta}) + \overline{P}(e^{i\lambda}, e^{i\theta})} = \left| e^{P(e^{i\lambda}, e^{i\theta})} \right|^2.$$

We note that $e^{P(e^{i\lambda}, e^{i\theta})} = 1 + Q(e^{i\lambda}, e^{i\theta})$, where Q is a continuous function with $\hat{Q}(m,n) = 0$ for all $(m,n) \notin S_0$. Hence, it follows from our above lemma that

$$\exp\left\{ \int_{[-\pi,\pi)} \int_{[-\pi,\pi)} \log w(e^{i\lambda}, e^{i\theta}) \, d\sigma^2(e^{i\lambda}, e^{i\theta}) \right\}$$

$$\geq \inf_P \int_{[-\pi,\pi)} \int_{[-\pi,\pi)} \left| 1 + P(e^{i\lambda}, e^{i\theta}) \right|^2 w(e^{i\lambda}, e^{i\theta}) \, d\sigma^2(e^{i\lambda}, e^{i\theta}),$$

where the infimum is taken over all continuous functions P with $\hat{P}(m,n) = 0$ for all $(m,n) \notin S_0$. The infimum will not increase if P is restricted to trigonometric polynomials with $\hat{P}(m,n) = 0$ for all $(m,n) \notin S_0$. Therefore, we have

$$\exp\left\{ \int_{[-\pi,\pi)} \int_{[-\pi,\pi)} \log w(e^{i\lambda}, e^{i\theta}) \, d\sigma^2(e^{i\lambda}, e^{i\theta}) \right\}$$

$$\geq \inf_P \int_{[-\pi,\pi)} \int_{[-\pi,\pi)} \left| 1 + P(e^{i\lambda}, e^{i\theta}) \right|^2 w(e^{i\lambda}, e^{i\theta}) \, d\sigma^2(e^{i\lambda}, e^{i\theta}),$$

where the infimum is taken over all trigonometric polynomials P with $\hat{P}(m,n) = 0$ for all $(m,n) \notin S_0$. It remains to show that this inequality is actually an equality. To do this, we start by taking $w = |1 + Q|^2$, where Q is a trigonometric polynomial with $\hat{Q}(m,n) = 0$ for all $(m,n) \notin S_0$. We then have

$$\exp\left\{ \int_{[-\pi,\pi)} \int_{[-\pi,\pi)} \log |1 + Q(e^{i\lambda}, e^{i\theta})|^2 \, d\sigma^2(e^{i\lambda}, e^{i\theta}) \right\}$$

$$\geq \inf_P \int_{[-\pi,\pi)} \int_{[-\pi,\pi)} \left| 1 + P(e^{i\lambda}, e^{i\theta}) \right|^2 |1 + Q(e^{i\lambda}, e^{i\theta})|^2 \, d\sigma^2(e^{i\lambda}, e^{i\theta})$$

$$= \inf_P \int_{[-\pi,\pi)} \int_{[-\pi,\pi)} \left| 1 + P(e^{i\lambda}, e^{i\theta}) + Q(e^{i\lambda}, e^{i\theta}) \right.$$

$$+ P(e^{i\lambda}, e^{i\theta}) Q(e^{i\lambda}, e^{i\theta})|^2 \, d\sigma^2(e^{i\lambda}, e^{i\theta}) \geq 1,$$

where the infimum is taken over all trigonometric polynomials P with $\hat{P}(m,n) = 0$ for all $(m,n) \notin S_0$. Hence, $\log |1 + Q|^2$ is in $L^1(\mathbb{T}^2)$ and

$$\int_{[-\pi,\pi)} \int_{[-\pi,\pi)} \log |1 + Q(e^{i\lambda}, e^{i\theta})|^2 \, d\sigma^2(e^{i\lambda}, e^{i\theta}) \geq 0.$$

Let $k = \exp\left\{ \int_{[-\pi,\pi)} \int_{[-\pi,\pi)} \log|1 + Q(e^{i\lambda}, e^{i\theta})|^2 \, d\sigma^2(e^{i\lambda}, e^{i\theta}) \right\}$. Then, $k \geq 1$ and if we set $\varphi = \log|1 + Q|^2 - \log k$, we see that φ is a real-valued function in $L^1(\mathbb{T}^2)$, with $\int_{[-\pi,\pi)} \int_{[-\pi,\pi)} \varphi(e^{i\lambda}, e^{i\theta}) \, d\sigma^2(e^{i\lambda}, e^{i\theta}) = 0$. Therefore, $|1 + Q|^2 = ke^{\varphi}$ and if we go back to our original w, we have

$$\int_{[-\pi,\pi)} \int_{[-\pi,\pi)} |1 + Q(e^{i\lambda}, e^{i\theta})|^2 \, w(e^{i\lambda}, e^{i\theta}) \, d\sigma^2(e^{i\lambda}, e^{i\theta})$$

$$= k \int_{[-\pi,\pi)} \int_{[-\pi,\pi)} e^{\varphi(e^{i\lambda}, e^{i\theta})} w(e^{i\lambda}, e^{i\theta}) \, d\sigma^2(e^{i\lambda}, e^{i\theta})$$

$$\geq \int_{[-\pi,\pi)} \int_{[-\pi,\pi)} e^{\varphi(e^{i\lambda}, e^{i\theta})} w(e^{i\lambda}, e^{i\theta}) \, d\sigma^2(e^{i\lambda}, e^{i\theta})$$

$$\geq \inf_{\psi} \int_{[-\pi,\pi)} \int_{[-\pi,\pi)} e^{\psi(e^{i\lambda}, e^{i\theta})} w(e^{i\lambda}, e^{i\theta}) \, d\sigma^2(e^{i\lambda}, e^{i\theta})$$

$$= \exp\left\{ \int_{[-\pi,\pi)} \int_{[-\pi,\pi)} \log w(e^{i\lambda}, e^{i\theta}) \, d\sigma^2(e^{i\lambda}, e^{i\theta}) \right\}.$$

Now, taking the infimum over all trigonometric polynomials Q with $\hat{Q}(m, n) = 0$ for all $(m, n) \notin S_0$, we get the opposite inequality and hence,

$$\exp\left\{ \int_{[-\pi,\pi)} \int_{[-\pi,\pi)} \log w(e^{i\lambda}, e^{i\theta}) \, d\sigma^2(e^{i\lambda}, e^{i\theta}) \right\}$$

$$= \inf_{P} \int_{[-\pi,\pi)} \int_{[-\pi,\pi)} |1 + P(e^{i\lambda}, e^{i\theta})|^2 \, w(e^{i\lambda}, e^{i\theta}) \, d\sigma^2(e^{i\lambda}, e^{i\theta}),$$

where the infimum is taken over all trigonometric polynomials P with $\hat{P}(m, n) = 0$ for all $(m, n) \notin S_0$. $\quad\square$

Theorem 2.10.2. *Let $f \in L^1(\mathbb{T}^2, \sigma^2)$ with Fourier series*

$$f(e^{i\lambda}, e^{i\theta}) \sim \sum_{S} b_{m,n} \, e^{-i(m\lambda + n\theta)},$$

where S is any semigroup of $\mathbb{Z}^2$. Then,

$$\int_{[-\pi,\pi)} \int_{[-\pi,\pi)} \log|f(e^{i\lambda}, e^{i\theta})| \, d\sigma^2(e^{i\lambda}, e^{i\theta}) \geq \log|b_{0,0}|. \qquad (2.20)$$

Proof. By Theorem 2.10.1,

$$
\exp\left\{ \int_{[-\pi,\pi)} \int_{[-\pi,\pi)} \log|f(e^{i\lambda}, e^{i\theta})| \, d\sigma^2(e^{i\lambda}, e^{i\theta}) \right\}
$$

$$
= \inf_P \int_{[-\pi,\pi)} \int_{[-\pi,\pi)} \left|1 + P(e^{i\lambda}, e^{i\theta})\right|^2 |f(e^{i\lambda}, e^{i\theta})| \, d\sigma^2(e^{i\lambda}, e^{i\theta}),
$$

where the infimum is taken over all trigonometric polynomials P with $\hat{P}(m,n) = 0$ for all $(m,n) \notin S_0$. If f is in $L^2(\mathbb{T}^2, \sigma^2)$, we can replace $|f|$, in the last formula, with $|f|^2$. Then, taking the square root of both sides, we get

$$
\exp\left\{ \int_{[-\pi,\pi)} \int_{[-\pi,\pi)} \log|f(e^{i\lambda}, e^{i\theta})| \, d\sigma^2(e^{i\lambda}, e^{i\theta}) \right\}
$$

$$
= \inf_P \left[\int_{[-\pi,\pi)} \int_{[-\pi,\pi)} \left|(1 + P(e^{i\lambda}, e^{i\theta}))f(e^{i\lambda}, e^{i\theta})\right|^2 \, d\sigma^2(e^{i\lambda}, e^{i\theta}) \right]^{1/2}. \quad (2.21)
$$

We now observe that the constant term of the Fourier series for $(1 + P)f$ is $b_{0,0}$. By the Parseval equality, the right side of (2.21) is at least $|b_{0,0}|$. Therefore, (2.20) holds.

If f is not in $L^2(\mathbb{T}^2, \sigma^2)$, let $\{f_n\}$ be the Fejér means of f. Each f_n is a trigonometric polynomial with constant term $b_{0,0}$, and the sequence converges to f in $L^1(\mathbb{T}^2)$. For every $\epsilon > 0$, and each n, we have

$$
\int_{[-\pi,\pi)} \int_{[-\pi,\pi)} \log(|f_n(e^{i\lambda}, e^{i\theta})| + \epsilon) \, d\sigma^2(e^{i\lambda}, e^{i\theta})
$$

$$
\geq \int_{[-\pi,\pi)} \int_{[-\pi,\pi)} \log|f_n(e^{i\lambda}, e^{i\theta})| \, d\sigma^2(e^{i\lambda}, e^{i\theta}) \geq \log|b_{0,0}|.
$$

With ϵ fixed and letting n go to infinity, we get

$$
\int_{[-\pi,\pi)} \int_{[-\pi,\pi)} \log(|f(e^{i\lambda}, e^{i\theta})| + \epsilon) \, d\sigma^2(e^{i\lambda}, e^{i\theta}) \geq \log|b_{0,0}|.
$$

We get our desired result by letting ϵ go to zero.   □

The following corollary follows immediately from this theorem.

Corollary 2.10.1. *If $f \in L^1(\mathbb{T}^2, \sigma^2)$ with mean value different from zero and Fourier series*

$$
f(e^{i\lambda}, e^{i\theta}) \sim \sum_S b_{m,n} \, e^{-i(m\lambda + n\theta)},
$$

where S is any semigroup of $\mathbb{Z}^2$, then $\log|f| \in L^1(\mathbb{T}^2, \sigma^2)$.

Theorem 2.10.3. *Let w be nonnegative and in $L^1(\mathbb{T}^2, \sigma^2)$, and let S be a semigroup of $\mathbb{Z}^2$. Then, w has a representation*

$$w(e^{i\lambda}, e^{i\theta}) = \left| \sum_{(m,n) \in S} b_{m,n} e^{-i(m\lambda + n\theta)} \right|^2,$$

with $b_{0,0} \neq 0$ and $\displaystyle\sum_{(m,n) \in S} |b_{m,n}|^2 < \infty$ if and only if

$$\int_{[-\pi,\pi)} \int_{[-\pi,\pi)} \log w(e^{i\lambda}, e^{i\theta}) \, d\sigma^2(e^{i\lambda}, e^{i\theta}) > -\infty.$$

Proof. ($\Rightarrow$) Let

$$f(e^{i\lambda}, e^{i\theta}) = \sum_{(m,n) \in S} b_{m,n} e^{-i(m\lambda + n\theta)}.$$

Then, $f \in L^2(\mathbb{T}^2, \sigma^2)$ since $\displaystyle\sum_{(m,n) \in S} |b_{m,n}|^2 < \infty$. Hence, $f \in L^1(\mathbb{T}^2, \sigma^2)$. Therefore, by Theorem 2.10.2, we have that

$$\int_{[-\pi,\pi)} \int_{[-\pi,\pi)} \log |f(e^{i\lambda}, e^{i\theta})| \, d\sigma^2(e^{i\lambda}, e^{i\theta}) \geq \log |b_{0,0}|,$$

and since $b_{0,0} \neq 0$, we have that $\log |b_{0,0}| > -\infty$. From these observations, we get

$$\int_{[-\pi,\pi)} \int_{[-\pi,\pi)} \log w(e^{i\lambda}, e^{i\theta}) \, d\sigma^2(e^{i\lambda}, e^{i\theta})$$

$$= \int_{[-\pi,\pi)} \int_{[-\pi,\pi)} \log |f(e^{i\lambda}, e^{i\theta})|^2 \, d\sigma^2(e^{i\lambda}, e^{i\theta})$$

$$= 2 \int_{[-\pi,\pi)} \int_{[-\pi,\pi)} \log |f(e^{i\lambda}, e^{i\theta})| \, d\sigma^2(e^{i\lambda}, e^{i\theta}) \geq 2 \log |b_{0,0}| > -\infty.$$

($\Leftarrow$) Let $\gamma = \displaystyle\int_{[-\pi,\pi)} \int_{[-\pi,\pi)} \log w(e^{i\lambda}, e^{i\theta}) \, d\sigma^2(e^{i\lambda}, e^{i\theta}) > -\infty$. Then, it was shown earlier that there exists a unique $H \in H^2_{S_0}(\mathbb{T}^2, \sigma^2)$ such that

$$e^{\gamma} = \int_{[-\pi,\pi)} \int_{[-\pi,\pi)} \left| 1 + H(e^{i\lambda}, e^{i\theta}) \right|^2 w(e^{i\lambda}, e^{i\theta}) \, d\sigma^2(e^{i\lambda}, e^{i\theta}). \qquad (2.22)$$

By Equation (2.16) and the surrounding remarks, it follows that

$$\left| 1 + H(e^{i\lambda}, e^{i\theta}) \right|^2 w(e^{i\lambda}, e^{i\theta})$$

is constant. Hence,

$$e^{\gamma} = \left| 1 + H(e^{i\lambda}, e^{i\theta}) \right|^2 w(e^{i\lambda}, e^{i\theta}).$$

Rewriting this equation, we get

$$w(e^{i\lambda}, e^{i\theta}) = \left| \frac{e^{\gamma/2}}{1 + H(e^{i\lambda}, e^{i\theta})} \right|^2 .$$

Since $w \in L^1(\mathbb{T}^2, \sigma^2)$, it follows that $\left(1 + H(e^{i\lambda}, e^{i\theta})\right)^{-1} \in L^2(\mathbb{T}^2, \sigma^2)$. To see that $\left(1 + H(e^{i\lambda}, e^{i\theta})\right)^{-1} \in H_S^2(\mathbb{T}^2, \sigma^2)$, note that

$$\int_{[-\pi,\pi)} \int_{[-\pi,\pi)} e^{-i(m\lambda+n\theta)} \frac{1}{1 + H(e^{i\lambda}, e^{i\theta})} \, d\sigma^2(e^{i\lambda}, e^{i\theta})$$

$$= \int_{[-\pi,\pi)} \int_{[-\pi,\pi)} e^{-i(m\lambda+n\theta)} \frac{1 + \overline{H}(e^{i\lambda}, e^{i\theta})}{|1 + H(e^{i\lambda}, e^{i\theta})|^2} \, d\sigma^2(e^{i\lambda}, e^{i\theta})$$

$$= e^{-\gamma} \int_{[-\pi,\pi)} \int_{[-\pi,\pi)} e^{-i(m\lambda+n\theta)} \left(1 + \overline{H}(e^{i\lambda}, e^{i\theta})\right) w(e^{i\lambda}, e^{i\theta}) \, d\sigma^2(e^{i\lambda}, e^{i\theta}).$$

It follows from (2.17) that the last integral is zero for all $(m, n) \in S_0$. Hence, $\left(1 + H(e^{i\lambda}, e^{i\theta})\right)^{-1} \in H_S^2(\mathbb{T}^2, \sigma^2)$. Therefore,

$$e^{\gamma/2} \left(1 + H(e^{i\lambda}, e^{i\theta})\right)^{-1} = \sum_{(m,n)\in S} b_{m,n} e^{-i(m\lambda+n\theta)}.$$

It remains to verify that $b_{0,0} \neq 0$. If $b_{0,0} = 0$, then

$$\int_{[-\pi,\pi)} \int_{[-\pi,\pi)} \left(1 + \overline{H}(e^{i\lambda}, e^{i\theta})\right) w(e^{i\lambda}, e^{i\theta}) \, d\sigma^2(e^{i\lambda}, e^{i\theta}) = 0.$$

It follows from this and the observations made above regarding (2.17) that

$$\int_{[-\pi,\pi)} \int_{[-\pi,\pi)} \left(1 + P(e^{i\lambda}, e^{i\theta})\right) \left(1 + \overline{H}(e^{i\lambda}, e^{i\theta})\right) w(e^{i\lambda}, e^{i\theta}) \, d\sigma^2(e^{i\lambda}, e^{i\theta}) = 0,$$

for every polynomial P in $H_{S_0}^2(\mathbb{T}^2, \sigma^2)$. If follows from this that

$$\int_{[-\pi,\pi)} \int_{[-\pi,\pi)} \left|1 + H(e^{i\lambda}, e^{i\theta})\right|^2 w(e^{i\lambda}, e^{i\theta}) \, d\sigma^2(e^{i\lambda}, e^{i\theta}) = 0,$$

contradicting (2.22). Therefore, $b_{0,0} \neq 0$ as desired.   □

Before we state our last theorem of the section, we introduce some notation and terminology. Let S be a semigroup and let $f \in H_S^2(\mathbb{T}^2, \sigma^2)$. Define $[f]_S$ be the smallest subspace of $H_S^2(\mathbb{T}^2, \sigma^2)$ that contains

$$\left(\sum_{(m,n)\in S'} a_{m,n} e^{-i(m\lambda+n\theta)} \right) f,$$

for all finite subsets S' of S. Finally, a function $f \in H_S^2(\mathbb{T}^2, \sigma^2)$ will be called outer if and only if

$$\int_{[-\pi,\pi)} \int_{[-\pi,\pi)} \log |f(e^{i\lambda}, e^{i\theta})| \, d\sigma^2(e^{i\lambda}, e^{i\theta})$$

$$= \log \left| \int_{[-\pi,\pi)} \int_{[-\pi,\pi)} f(e^{i\lambda}, e^{i\theta}) \, d\sigma^2(e^{i\lambda}, e^{i\theta}) \right| > -\infty.$$

Theorem 2.10.4. *$f \in H_S^2(\mathbb{T}^2, \sigma^2)$ is outer if and only if $[f]_S = H_S^2(\mathbb{T}^2, \sigma^2)$.*

Proof. Using the ideas developed in the proof of Theorem 2.10.2, in particular, Equation (2.21) and Parseval equality, we get that $f \in H_S^2(\mathbb{T}^2, \sigma^2)$ being outer is equivalent to the existence of a sequence of polynomials P_n in $H_{S_0}^2(\mathbb{T}^2, \sigma^2)$ such that $(1 + P_n)f$ converges to

$$\int_{[-\pi,\pi)} \int_{[-\pi,\pi)} f(e^{i\lambda}, e^{i\theta}) \, d\sigma^2(e^{i\lambda}, e^{i\theta}),$$

a constant function. This in turn is equivalent to $[f]_S = H_S^2(\mathbb{T}^2, \sigma^2)$. $\qquad\square$

2.11 Semigroup Moving Average Representations

Recall that a subset S of $\mathbb{Z}^2$ is a semigroup if:

(a) $S \cup (-S) = \mathbb{Z}^2$,

(b) $S \cap (-S) = \{(0,0)\}$, and

(c) if (m, n) and (m', n') are in S, then $(m + m', n + n')$ is in S.

Let $(\Omega, \mathcal{F}, P)$ be a probability space. Let $X_{m,n}$, $(m, n) \in \mathbb{Z}^2$ be a random field. Suppose further that $X_{m,n}$, $(m, n) \in \mathbb{Z}^2$ is an S-semigroup moving average random field. That is, there exists a sequence of complex numbers $(a_{m,n})_{(m,n) \in S}$ with the property that $\sum_{(m,n) \in S} |a_{m,n}|^2 < \infty$ and a white noise random field $\xi_{m,n}$, $(m, n) \in \mathbb{Z}^2$ in $L^2(\Omega, \mathcal{F}, P)$ such that

$$X_{m,n} = \sum_{(j,k) \in S} a_{j,k} \, \xi_{m-j,n-k},$$

in $L^2(\Omega, \mathcal{F}, P)$, for all $(m, n) \in \mathbb{Z}^2$. A straightforward calculation shows that $X_{m,n}$, $(m, n) \in \mathbb{Z}^2$ is a weakly stationary random field.

We begin by examining the spectral measure for this random field. First recall that white noise is a stationary random field and as such has a spectral representation.

$$\xi_{m,n} = \int_{[-\pi,\pi)} \int_{[-\pi,\pi)} e^{im\lambda + in\theta} dZ_\xi(e^{i\lambda}, e^{i\theta}),$$

where Z_ξ is the orthogonally scattered set function associated with the random field $\xi_{m,n}$, $(m,n) \in \mathbb{Z}^2$. It is straightforward to verify that F_ξ, the spectral measure for $\xi_{m,n}$, $(m,n) \in \mathbb{Z}^2$, is normalized Lebesgue measure on $\mathbb{T}^2$. That is, $dF_\xi(e^{i\lambda}, e^{i\theta}) = d\sigma^2(e^{i\lambda}, e^{i\theta})$. We now calculate the covariance function for $X_{m,n}$, $(m,n) \in \mathbb{Z}^2$.

$$
\begin{aligned}
r_X(m,n) &= (X_{m,n}, X_{0,0})_{L^2(\Omega,\mathcal{F},P)} \\
&= \left(\sum_{(j,k)\in S} a_{j,k}\,\xi_{m-j,n-k}, \sum_{(j,k)\in S} a_{j,k}\,\xi_{-j,-k} \right)_{L^2(\Omega,\mathcal{F},P)} \\
&= \int_{[-\pi,\pi)} \int_{[-\pi,\pi)} \sum_{(j,k)\in S} a_{j,k} e^{i(m-j)\lambda + i(n-k)\theta} \\
&\quad \times \overline{\sum_{(j,k)\in S} a_{j,k} e^{-ij\lambda - ik\theta}}\; dF_\xi(e^{i\lambda}, e^{i\theta}) \\
&= \int_{[-\pi,\pi)} \int_{[-\pi,\pi)} \left| \sum_{(j,k)\in S} a_{j,k} e^{-ij\lambda - ik\theta} \right|^2 e^{im\lambda + in\theta}\, d\sigma^2(e^{i\lambda}, e^{i\theta}).
\end{aligned}
$$

Therefore, we see that the spectral measure of $X_{m,n}$, $(m,n) \in \mathbb{Z}^2$, is absolutely continuous with respect to normalized Lebesgue measure on $\mathbb{T}^2$, with density

$$f_X(e^{i\lambda}, e^{i\theta}) = \left| \sum_{(j,k)\in S} a_{j,k}\, e^{-ij\lambda - ik\theta} \right|^2 = \left| \sum_{(j,k)\in S} \overline{a_{j,k}}\, e^{ij\lambda + ik\theta} \right|^2.$$ We now record

this observation as a theorem.

Theorem 2.11.1. *Suppose that* $X_{m,n}$, $(m,n) \in \mathbb{Z}^2$ *is an S-semigroup moving average random field; that is, there exists a sequence of complex numbers* $(a_{m,n})_{(m,n)\in S}$ *with the property that* $\displaystyle\sum_{(m,n)\in S} |a_{m,n}|^2 < \infty$ *and a white noise random field* $\xi_{m,n}$, $(m,n) \in \mathbb{Z}^2$ *in* $L^2(\Omega, \mathcal{F}, P)$ *such that*

$$X_{m,n} = \sum_{(j,k)\in S} a_{j,k}\,\xi_{m-j,n-k},$$

in $L^2(\Omega, \mathcal{F}, P)$, *for all* $(m,n) \in \mathbb{Z}^2$. *Then,* $X_{m,n}$, $(m,n) \in \mathbb{Z}^2$ *is a weakly stationary random field that has a spectral measure that is absolutely continuous with respect to normalized Lebesgue measure with density* $f_X(e^{i\lambda}, e^{i\theta}) = |\varphi(e^{i\lambda}, e^{i\theta})|^2$, *where* $\varphi \in L^2(\mathbb{T}^2, \sigma^2)$, $\hat{\varphi}(j,k) = \overline{a_{j,k}}$ *for all* $(j,k) \in S$ *and* $\hat{\varphi}(j,k) = 0$ *for all* $(j,k) \in \mathbb{Z}^2 \setminus S$.

Now, let us examine the situation from the other direction. Suppose that $X_{m,n}$, $(m,n) \in \mathbb{Z}^2$ is a weakly stationary random field with spectral measure $dF_X(e^{i\lambda}, e^{i\theta}) = |\varphi(e^{i\lambda}, e^{i\theta})|^2 d\sigma^2(e^{i\lambda}, e^{i\theta})$, where $\varphi \in L^2(\mathbb{T}^2, \sigma^2)$, with Fourier series of the form $\sum\limits_{(j,k) \in S} \hat{\varphi}(j,k) e^{ij\lambda + ik\theta}$. As such, φ is equal to its Fourier series in $L^2(\mathbb{T}^2, \sigma^2)$. That is,

$$\varphi(e^{i\lambda}, e^{i\theta}) = \sum_{(j,k) \in S} \hat{\varphi}(j,k) e^{ij\lambda + ik\theta} \text{ in } L^2(\mathbb{T}^2, \sigma^2).$$

We now define the linear transformation $W : L(X) \to L^2(\mathbb{T}^2, \sigma^2)$ by $W(X_{m,n}) = e^{im\lambda + in\theta} \overline{\varphi}(e^{i\lambda}, e^{i\theta})$. W is an isometry and as such, we may identify $L(X)$ and $W(L(X))$, as well as, $X_{m,n}$ and $e^{im\lambda + in\theta} \overline{\varphi}(e^{i\lambda}, e^{i\theta})$. $e^{im\lambda + in\theta} \overline{\varphi}(e^{i\lambda}, e^{i\theta})$ has a moving average representation since

$$e^{im\lambda + in\theta} \overline{\varphi}(e^{i\lambda}, e^{i\theta}) = \sum_{(j,k) \in S} \overline{\hat{\varphi}}(j,k) e^{i(m-j)\lambda + i(n-k)\theta} \text{ in } L^2(\mathbb{T}^2, \sigma^2).$$

So, we may conclude the $X_{m,n}$, $(m,n) \in \mathbb{Z}^2$ has a moving average representation of the form

$$X_{m,n} = \sum_{(j,k) \in S} \overline{\hat{\varphi}(j,k)} \xi_{m-j,n-k},$$

for some white noise random field $(\xi_{m,n})_{(m,n) \in \mathbb{Z}^2}$ contained in some Hilbert space that contains $L(X)$ as a subspace. A natural question at this point might be: when is this white noise sequence contained in $L(X)$? Let us examine that question. Let

$$L(\xi) = \overline{\text{span}} \left\{ \xi_{m,n} : (m,n) \in \mathbb{Z}^2 \right\}.$$

Now, define the linear transformation $W^\# : L(\xi) \to L^2(\mathbb{T}^2, \sigma^2)$ by $W^\#(\xi_{m,n}) = e^{im\lambda + in\theta}$. $W^\#$ is clearly an isomorphism. It follows from the moving average representation of $X_{m,n}$, $(m,n) \in \mathbb{Z}^2$, that $L(X) \subseteq L(\xi)$. Again from this moving average representation, we see that

$$W^\#(X_{m,n}) = e^{im\lambda + in\theta} \overline{\varphi}(e^{i\lambda}, e^{i\theta}).$$

It follows then that

$$W^\#(L(X)) = \overline{\text{span}} \left\{ e^{im\lambda + in\theta} \overline{\varphi}(e^{i\lambda}, e^{i\theta}) : (m,n) \in \mathbb{Z}^2 \right\},$$

which is a doubly invariant subspace of $L^2(\mathbb{T}^2, \sigma^2)$. By Theorem 3.6.1, all doubly invariant subspaces of $L^2(\mathbb{T}^2, \sigma^2)$ are of the form $\mathbf{1}_E L^2(\mathbb{T}^2, \sigma^2)$, where E is a Lebesgue measurable subset of $\mathbb{T}^2$. In our particular case, $E = \left\{ (e^{i\lambda}, e^{i\theta}) \in \mathbb{T}^2 : \varphi(e^{i\lambda}, e^{i\theta}) \neq 0 \right\}$. From this observation, we see that if E^c, the complement of E, has Lebesgue measure zero, then $W^\#(L(X)) = L^2(\mathbb{T}^2, \sigma^2)$. This together with the fact that $W^\#$ is an isomorphism gives us that $L(X) = L(\xi)$ and therefore, $\xi_{m,n} \in L(X)$ for all $(m,n) \in \mathbb{Z}^2$. If, on the other hand, the Lebesgue measure of E^c is positive, then $L(X) \subsetneq L(\xi)$. So, there exists a $(j,k) \in \mathbb{Z}^2$ such that $\xi_{j,k} \notin L(X)$. An important observation to make here, if not already observed, is that $|\varphi(e^{i\lambda}, e^{i\theta})|^2$ is the density of the spectral measure of our random field $X_{m,n}$, $(m, n) \in \mathbb{Z}^2$. Therefore, whether or not our spectral density vanishes on a set of positive measure impacts the location of our white noise random field.

We will now record these observations as a theorem.

Theorem 2.11.2. *Suppose that $X_{m,n}$, $(m, n) \in \mathbb{Z}^2$ is a weakly stationary random field with spectral measure $dF_X(e^{i\lambda}, e^{i\theta}) = |\varphi(e^{i\lambda}, e^{i\theta})|^2 \, d\sigma^2(e^{i\lambda}, e^{i\theta})$, where $\varphi \in L^2(\mathbb{T}^2, \sigma^2)$, with Fourier series of the form $\sum\limits_{(j,k)\in S} \hat{\varphi}(j, k)e^{ij\lambda + ik\theta}$.*
Then, $X_{m,n}$, $(m, n) \in \mathbb{Z}^2$ may be represented as a moving average random field; that is,

$$X_{m,n} = \sum_{(j,k)\in S} \overline{\hat{\varphi}(j, k)}\xi_{m-j,n-k},$$

in some Hilbert space $\mathcal{H}$ containing the white noise sequence $\xi_{m,n}$, $(m, n) \in \mathbb{Z}^2$ and having $L(X)$ as a subspace. If $\varphi \neq 0$ $[\sigma^2]$-a.e., then our white noise sequence is contained in $L(X)$.

We now examine the proper moving averages associated with a semigroup. First, we note that every semigroup induces an ordering on $\mathbb{Z}^2$ as follows:

$$(j, k) \prec (m, n) \text{ if and only if } (m - j, n - k) \in S.$$

With such an ordering, we define

$$L(X : (m, n)) = \overline{\text{span}}\left\{X_{j,k} : (j, k) \prec (m, n)\right\}.$$

Suppose that $X_{m,n}$, $(m, n) \in \mathbb{Z}^2$ is an S-semigroup moving average random field. Therefore, there exists a sequence of complex numbers $(a_{m,n})_{(m,n)\in S}$ with the property that $\sum\limits_{(m,n)\in S} |a_{m,n}|^2 < \infty$ and a white noise random field $\xi_{m,n}$, $(m, n) \in \mathbb{Z}^2$ in $L^2(\Omega, \mathcal{F}, P)$ such that

$$X_{m,n} = \sum_{(j,k)\in S} a_{j,k}\, \xi_{m-j,n-k},$$

in $L^2(\Omega, \mathcal{F}, P)$, for all $(m, n) \in \mathbb{Z}^2$. Such a moving average is called **proper** if

$$L(X : (m, n)) = L(\xi : (m, n)), \text{ for all } (m, n) \in \mathbb{Z}^2.$$

From the moving average representation, we see that

$$W^{\#}(X_{m,n}) = e^{im\lambda + in\theta} \sum_{(j,k)\in S} a_{j,k}e^{-ij\lambda - ik\theta},$$

where $W^{\#}$ is defined above. It follows from this calculation that

$$
\begin{aligned}
W^{\#}(L(X : (m, n))) &= \overline{\text{span}}\left\{e^{ij\lambda + ik\theta} f(e^{i\lambda}, e^{i\theta}) : (j, k) \prec (m, n)\right\} \\
&= e^{im\lambda + in\theta}[f]_S,
\end{aligned}
$$

where $f(e^{i\lambda}, e^{i\theta}) = \sum\limits_{(m,n)\in S} a_{m,n} e^{-im\lambda - in\theta}$ and

$$W^{\#}(L(\xi : (m,n))) = \overline{\text{span}}\left\{ e^{ij\lambda + ik\theta} : (j,k) \prec (m,n) \right\} = e^{im\lambda + in\theta} H_S^2(\mathbb{T}^2, \sigma^2).$$

It then follows from Theorem 2.10.4 that $L(X : (m,n)) = L(\xi : (m,n))$, for all $(m,n) \in \mathbb{Z}^2$ if and only if f is outer. We summarize this in the following theorem.

Theorem 2.11.3. *Suppose that $X_{m,n}$, $(m,n) \in \mathbb{Z}^2$ is an S-semigroup moving average random field. Therefore, there exists a sequence of complex numbers $(a_{m,n})_{(m,n)\in S}$ with the property that $\sum\limits_{(m,n)\in S} |a_{m,n}|^2 < \infty$ and a white noise random field $\xi_{m,n}$, $(m,n) \in \mathbb{Z}^2$ in $L^2(\Omega, \mathcal{F}, P)$ such that*

$$X_{m,n} = \sum_{(j,k)\in S} a_{j,k}\, \xi_{m-j,n-k},$$

in $L^2(\Omega, \mathcal{F}, P)$, for all $(m,n) \in \mathbb{Z}^2$. This moving average representation is proper if and only if f is outer, where $f(e^{i\lambda}, e^{i\theta}) = \sum\limits_{(m,n)\in S} a_{m,n} e^{-im\lambda - in\theta}$.

2.12 Wold-Type Decompositions

Every semigroup S induces an ordering on $\mathbb{Z}^2$ as follows:

$$(m', n') \stackrel{S}{\prec} (m, n) \text{ if and only if } (m' - m, n' - n) \in S.$$

A particular semigroup, which has been found useful in practice, is the so-called **asymmetric horizontal half-space** S given by

$$S = \left\{ (m,n) \in \mathbb{Z}^2 : m \leq -1, n \in \mathbb{Z} \right\} \cup \left\{ (0,n) : n \leq 0 \right\}.$$

Throughout this section, we shall be using this particular semigroup.

Let $X_{m,n}$, $(m,n) \in \mathbb{Z}^2$ be a weakly stationary random field with ordering on $\mathbb{Z}^2$ induced by the semigroup S. We define the past of the random field up to (m,n) to be

$$L^S(X : (m,n)) = \overline{\text{span}}\left\{ X_{j,k} : (j,k) \stackrel{S}{\prec} (m,n) \right\}.$$

We note that for each $m \in \mathbb{Z}$,

$$L^1(X : m - 1) \subseteq L^S(X : (m, n-1)) \subseteq L^1(X : m). \tag{2.23}$$

We define the S-innovation at (m,n) by

$$\nu_{m,n}^S = X_{m,n} - P_{L^S(X : (m, n-1))} X_{m,n}.$$

We now give the following definitions, first introduced in [10].

Definition 2.12.1. *A weakly stationary random field,* $X_{m,n}$, $(m,n) \in \mathbb{Z}^2$, *is called*

(a) *S-deterministic if* $L^S(X:(m,n)) = L(X)$ *for all* $(m,n) \in \mathbb{Z}^2$,

(b) *S-innovation if* $L(\nu^S) = L(X)$, *and*

(c) *S-evanescent if* $L(X) = [L^1(X:-\infty) \oplus L(\nu^S)]^\perp$.

By (2.23), we get the following lemma.

Lemma 2.12.1. *A weakly stationary random field is S-deterministic if and only if it is horizontally singular.*

We observe that

$$L^S(X:(m,n-1)) = L^1(X:m-1) \oplus L(\eta^m:n-1), \qquad (2.24)$$

where $\eta_n^m = \nu_{m,n}^1$, for $n \in \mathbb{Z}$ and m fixed with

$$L(\eta^m:n-1) = \overline{span}\,\{\eta_k^m:k \le n-1\}$$

and

$$L^1(X:m) = L^1(X:m-1) \oplus L(\eta^m),$$

for each $m \in \mathbb{Z}$, where $L(\eta^m) = \overline{span}\,\{\eta_k^m:k \in \mathbb{Z}\}$.

Before we state our next lemma, recall that $\nu_{m,n}^1 = X_{m,n} - P_{L^1(X:m-1)}X_{m,n}$.

Lemma 2.12.2. $\nu_{m,n}^S = \nu_{m,n}^1 - P_{L(\eta^m:n-1)}\nu_{m,n}^1$.

Proof. By definition, we have that $\nu_{m,n}^S = X_{m,n} - P_{L^S(X:(m,n-1))}X_{m,n}$. By Equation (2.24), it follows that $P_{L^S(X:(m,n-1))}X_{m,n} = P_{L^1(X:m-1)}X_{m,n} + P_{L(\eta^m:n-1)}X_{m,n}$. Therefore, it follows that $\nu_{m,n}^S = X_{m,n} - P_{L^1(X:m-1)}X_{m,n} - P_{L(\eta^m:n-1)}X_{m,n}$. Hence, $\nu_{m,n}^S = \nu_{m,n}^1 - P_{L(\eta^m:n-1)}\nu_{m,n}^1$, since $P_{L(\eta^m:n-1)}X_{m,n} = P_{L(\eta^m:n-1)}\nu_{m,n}^1$, which follows from the fact that $P_{L(\eta^m:n-1)}P_{L^1(X:m-1)}X_{m,n} = 0$. $\qquad \square$

The following theorem is an explicit form of a decomposition, given in [10], for weakly stationary random fields.

Theorem 2.12.1. *Every weakly stationary random field,* $X_{m,n}$, $(m,n) \in \mathbb{Z}^2$, *has the following unique decomposition*

$$X_{m,n} = X_{m,n}^{(i)} + X_{m,n}^{(e)} + X_{m,n}^{(d)},$$

where

(a) $X_{m,n}^{(i)}$ *is S-innovation,*

(b) $X_{m,n}^{(e)}$ is S-evanescent,

(c) $X_{m,n}^{(d)}$ is S-deterministic, and

(d) $L(X^{(i)}), L(X^{(e)}), L(X^{(d)}) \subseteq L(X)$ are orthogonal subspaces.

Proof. First, observe that

$$L^S(X:(m,n)) = \left(\sum_{k=1}^{\infty} \oplus \left[L^1(X:m-k) \ominus L^1(X:m-k-1) \right] \right)$$
$$\oplus \quad L^1(X:-\infty) \oplus L(\eta^m:n)$$

For m fixed, we may apply the Wold Decomposition to η_n^m to get $\eta_n^m = \eta_n^{m,r} + \eta_n^{m,s}$. We then get

$$L^1(X:m-k) \ominus L^1(X:m-k-1) = \overline{span}\left\{ \eta_n^{m-k} : n \in \mathbb{Z} \right\}$$

$$= \overline{span}\left\{ \eta_n^{m-k,r} : n \in \mathbb{Z} \right\} \oplus \overline{span}\left\{ \eta_n^{m-k,s} : n \in \mathbb{Z} \right\}.$$

This together with the fact that $\eta_n^{m,s}$ is singular yields

$$L(\eta^m:n) = L(\eta^{m,r}:n) \oplus L(\eta^{m,s}).$$

Thus,

$$L^S(X:(m,n)) = \left(\sum_{k=1}^{\infty} \oplus \left[L(\eta^{m-k,r}) \oplus L(\eta^{m-k,s}) \right] \right)$$
$$\oplus \quad L^1(X:-\infty) \oplus L(\eta^{m,r}:n) \oplus L(\eta^{m,s}).$$

Reorganizing, we get

$$L^S(X:(m,n)) = \left(L(\eta^{m,r}:n) \oplus \sum_{k=1}^{\infty} \oplus L(\eta^{m-k,r}) \right)$$
$$\oplus \left(\sum_{k=0}^{\infty} \oplus L(\eta^{m-k,s}) \right) \oplus L^1(X:-\infty).$$

If we now let $X_{m,n}^{(i)}$ be the projection of $X_{m,n}$ on the first part our decom-position, $X_{m,n}^{(e)}$ be the projection of $X_{m,n}$ on the second part our decompo-sition and $X_{m,n}^{(d)}$ be the projection of $X_{m,n}$ on the last part our decomposi-tion, we have the orthogonal decomposition $X_{m,n} = X_{m,n}^{(i)} + X_{m,n}^{(e)} + X_{m,n}^{(d)}$, with each piece having the appropriate properties. We recall that $\eta_n^{m,r}$, $(m,n) \in \mathbb{Z}^2$ is an orthogonal sequence. Further, if we let $\nu_{m,n} = \eta_n^{m,r}$ and $\xi_{m,m-k,n} = P_{L(\eta^{m-k,s})}X_{m,n}$, then we can write

$$X_{m,n}^{(i)} = \sum_{l=0}^{\infty} a_{0,l}\nu_{m,n-l} + \sum_{k=1}^{\infty}\sum_{l=-\infty}^{\infty} a_{k,l}\nu_{m-k,n-l}$$

and

$$X_{m,n}^{(e)} = \sum_{k=0}^{\infty} \xi_{m,m-k,n}.$$

□

Let $X_{m,n}^{(s)} = X_{m,n}^{(e)} + X_{m,n}^{(d)}$. We call $X^{(s)}$ the singular component of X. With this notation, we get the following Wold Decomposition with respect to the pasts $L^S(X:(m,n))$.

Theorem 2.12.2 (Wold Decomposition). *Every weakly stationary random field has a decomposition in the form*

$$X_{m,n} = X_{m,n}^{(i)} + X_{m,n}^{(s)},$$

where $L\left(X^{(i)}\right) \perp L\left(X^{(s)}\right)$ *and* $L^S(X:(m,n)) \supseteq L^S(X^{(i)}:(m,n))$.

Corollary 2.12.1. *Let* $X_{m,n}$, $(m,n) \in \mathbb{Z}^2$, *be a weakly stationary random field. Then*

(a) $X_{m,n} = X_{m,n}^{(s)}$ *if and only if*

$$\int_{[-\pi,\pi)} \int_{[-\pi,\pi)} \log\left(f_X(e^{i\lambda}, e^{i\theta})\right) d\sigma^2(e^{i\lambda}, e^{i\theta}) = -\infty,$$

where f_X *is the density of the absolutely continuous part of* F_X *with respect to* σ^2.

(b) $X_{m,n} = X_{m,n}^{(d)}$ *if and only if for* $[F_{X2}]$-*a.e.* θ,

$$\int_{[-\pi,\pi)} \log\left(g_a(e^{i\lambda}, e^{i\theta})\right) d\sigma(e^{i\lambda}) = -\infty,$$

where g_a *is the density of the absolutely continuous part of* F_X *with respect to* $\sigma \otimes F_{X2}$.

(c) $X_{m,n} = X_{m,n}^{(e)}$ *if and only if*

    (i) F_X *is absolutely continuous with respect to* $\sigma \otimes F_{X2}$,

    (ii) *for* $[F_{X2}]$-*a.e.* $e^{i\theta}$,

$$\int_{[-\pi,\pi)} \log\left(\frac{dF_X(e^{i\lambda}, e^{i\theta})}{d(\sigma \otimes F_2)}\right) d\sigma(e^{i\lambda}) > -\infty,$$

    and

    (iii)

$$\int_{[-\pi,\pi)} \int_{[-\pi,\pi)} \log\left(f_X(e^{i\lambda}, e^{i\theta})\right) d\sigma^2(e^{i\lambda}, e^{i\theta}) = -\infty,$$

where f_X *is the density of the absolutely continuous part of* F_X *with respect to* σ^2.

(d) $X_{m,n} = X_{m,n}^{(i)}$ if and only if

(i) F_X is absolutely continuous with respect to σ^2, and

(ii)
$$\int_{[-\pi,\pi)} \int_{[-\pi,\pi)} \log\left(\frac{dF_X(e^{i\lambda}, e^{i\theta})}{d\sigma^2}\right) d\sigma^2(e^{i\lambda}, e^{i\theta}) > -\infty.$$

Proof. (a): First, recall that $\nu_{m,n}^S = \eta_n^m - P_{L(\eta^m:n-1)}\eta_n^m$. Hence,

$$E\left|\nu_{m,n}^S\right|^2 = E\left|\eta_n^0 - P_{L(\eta^0:n-1)}\eta_n^0\right|^2.$$

Now, if we let $f_{\nu^1,2}$ denote the absolutely continuous part of the spectral measure of η_n^0 with respect to σ, then by the Szegö-Krein-Kolmogorov Theorem, it follows that

$$E\left|\eta_n^0 - P_{L(\eta^0:n-1)}\eta_n^0\right|^2 = \exp\left(\int_{[-\pi,\pi)} \log\left(f_{\nu^1,2}(e^{i\theta})\right) d\sigma(e^{i\theta})\right).$$

Using Equation (2.3), we get

$$E\left|\nu_{m,n}^S\right|^2 = \exp\left(\int_{[-\pi,\pi)} \int_{[-\pi,\pi)} \log\left(f_X(e^{i\lambda}, e^{i\theta})\right) d\sigma(e^{i\lambda}) d\sigma(e^{i\theta})\right).$$

From this equation, (a) follows.

(b): In light of Lemma 2.12.1, Theorem 2.7.1 and Theorem 2.7.3 together give us (b).

(c): It follows from the definition that if $X_{m,n} = X_{m,n}^{(e)}$, then $X_{m,n}$ is horizontally regular. Therefore, by Theorem 2.6.1, (i) and (ii) hold. It then follows from (a) that (iii) also holds. For the other direction, once again, by Theorem 2.6.1 if follows from (i) and (ii) that $X_{m,n}$ is horizontally regular. Again, using (a), we see that $X_{m,n}$ must be in the singular component. Therefore, $X_{m,n} = X_{m,n}^{(e)}$, as desired.

(d): The sufficiency of the conditions follows from Theorem 2.12.2 and part (a) of this corollary. For necessity of the conditions, note that if $X_{m,n} = X_{m,n}^{(i)}$, then by definition and Theorem 2.12.1, it follows that $X_{m,n}$ is horizontally regular and so by Theorem 2.6.1, F_X is absolutely continuous with respect to $\sigma \otimes F_{X2}$ and $\int_{[-\pi,\pi)} \log \frac{dF_X(e^{i\lambda}, e^{i\theta})}{d(\sigma \otimes F_{X2})} d\sigma(e^{i\lambda}) > -\infty$ for $[F_{X2}]$-a.e. $e^{i\theta}$. In addition, we have that $\{\eta_n^0\}$ is regular giving $F_{X2} \equiv F_{\nu^1,2}$, which is absolutely continuous with respect to σ. Therefore, F_X is absolutely continuous with respect to σ^2. Now, since $E|\nu^S|^2 \neq 0$, it follows that

$$\int_{[-\pi,\pi)} \int_{[-\pi,\pi)} \log\left(\frac{dF_X(e^{i\lambda}, e^{i\theta})}{d\sigma^2}\right) d\sigma^2(e^{i\lambda}, e^{i\theta}) > -\infty.$$

□

Theorem 2.12.3. *A random field, $X_{m,n}$, $(m,n) \in \mathbb{Z}^2$, is S-innovation if and only if F_X is absolutely continuous with respect to σ^2 and the spectral density $f(e^{i\lambda}, e^{i\theta}) = |\varphi(e^{i\lambda}, e^{i\theta})|^2$, where $\varphi \in L^2(\mathbb{T}^2, \sigma^2)$ with $\hat{\varphi}(j,k) = 0$ for all $(j,k) \in \mathbb{Z}^2 \setminus S$.*

Proof. We observed in the proof of Theorem 2.12.1 that if $X_{m,n}$ is an S-innovation, then $X_{m,n}$ can be written in the form

$$X_{m,n} = \sum_{l=0}^{\infty} a_{0,l} \nu_{m,n-l} + \sum_{k=1}^{\infty} \sum_{l=-\infty}^{\infty} a_{k,l} \nu_{m-k,n-l},$$

where $\{\nu_{m,n}\}$ is an orthonormal sequence. The desired conclusion follows from this S-semigroup moving average representation and Theorem 2.11.1. For the other direction, note that under these conditions, Theorem 2.10.3 and part (d) of the above corollary the desired conclusion follows.   □

Now, let

$$S_1 = \{(m,n) \in \mathbb{Z}^2 : m \leq -1, n \in \mathbb{Z}\} \cup \{(0,n) : n \leq 0\}$$

and

$$S_2 = \{(m,n) \in \mathbb{Z}^2 : m \in \mathbb{Z}, n \leq -1\} \cup \{(m,0) : m \leq 0\}.$$

Note, S_1 is S as defined at the beginning of the section. The following result is an immediate corollary of our last theorem.

Corollary 2.12.2. *A random field, $X_{m,n}$, $(m,n) \in \mathbb{Z}^2$, is S_1 and S_2-innovation if and only if F_X is absolutely continuous with respect to σ^2 and the spectral density $f(e^{i\lambda}, e^{i\theta}) = |\varphi(e^{i\lambda}, e^{i\theta})|^2$, where $\varphi \in H^2(\mathbb{T}^2)$.*

One can observe from this that $\hat{\varphi}(0,0) \neq 0$. We now make the following observation regarding $H^2(\mathbb{T}^2)$ functions, which follows from this observation and the above corollary and theorem.

Corollary 2.12.3. $\int_{[-\pi,\pi)} \int_{[-\pi,\pi)} \log(f(e^{i\lambda}, e^{i\theta}))\, d\sigma^2(e^{i\lambda}, e^{i\theta}) > -\infty$ *if and only if* $f(e^{i\lambda}, e^{i\theta}) = |\varphi(e^{i\lambda}, e^{i\theta})|^2$, *where* $\varphi \in H^2(\mathbb{T}^2)$ *and* $\hat{\varphi}(0,0) \neq 0$.

2.13 Remarks and Related Literature

There are several papers on stationary random fields. For special examples of stationary random fields one might consider [PW]. Spatial processes are important in studying Gibbs phenomenon for which the Markov property is important. For readers interested in these topics, one might consider [PM]. The only work related to the problems studied here is due to Tse-Pei Chiang [TPC], a student of Yaglom, who studied half-space ordering. As stated in the introduction, to study analytic aspects, one needs lexicographic ordering, which is studied here. It leads to more interesting spectral conditions.

[PW] P. Whittle, *On stationary processes in the plane.* Biometrika 41, (1954), 434–449.

[PM] P. A. P. Moran, *A Gaussian Markovian process on a square lattice.* J. Appl. Probability 10 (1973), 54–62.

[TPC] Tse-Pei Chiang, *Extrapolation theory of a homogeneous random field with continuous parameters.* (Russian. English summary) Teor. Veroyatnost. i Primenen. 2 (1957), 60–91.

3

Invariant Subspaces

3.1 The Halmos Decomposition

Let $\mathcal{H}$ be a Hilbert space and let $\mathcal{L}(\mathcal{H})$ denote the collection of all bounded linear operators on $\mathcal{H}$. An operator $T \in \mathcal{L}(\mathcal{H})$ is called an **isometry** if

$$(Tf, Tg)_{\mathcal{H}} = (f, g)_{\mathcal{H}}$$

for all $f, g \in \mathcal{H}$. An isometry S in $\mathcal{L}(\mathcal{H})$ is called a **shift operator** if

$$\|S^{*n} f\|_{\mathcal{H}} \to 0, \text{ as } n \to \infty,$$

for all $f \in \mathcal{H}$. An isometry T in $\mathcal{L}(\mathcal{H})$ is called a **unitary operator** if $T(\mathcal{H}) = \mathcal{H}$. If T is a unitary operator in $\mathcal{L}(\mathcal{H})$, then T^* is an isometry and $T^* = T^{-1}$. Therefore, T^* is also a unitary operator.

If T is an isometry in $\mathcal{L}(\mathcal{H})$, then

$$0 \leq \|T^{*n} f\|_{\mathcal{H}} \leq \|f\|_{\mathcal{H}}.$$

By definition, if T is a shift operator, then

$$\|T^{*n} f\|_{\mathcal{H}} \to 0, \text{ as } n \to \infty,$$

for all $f \in \mathcal{H}$ and if T is a unitary operator, then

$$\|T^{*n} f\|_{\mathcal{H}} = \|f\|_{\mathcal{H}}$$

for all $f \in \mathcal{H}$. These observations show us that these two subclasses of isometries are the two extreme cases. The Halmos Decomposition tells us that for every isometry in $\mathcal{L}(\mathcal{H})$, $\mathcal{H}$ may be decomposed into the (orthogonal) direct sum of two subspaces. On one subspace, T is a shift operator and on the other subspace, T is a unitary operator. Before we state the Halmos Decomposition in detail, we recall that a subspace $\mathcal{M}$ of $\mathcal{H}$ is said to **reduce** an isometry T in $\mathcal{L}(\mathcal{H})$, if $T(\mathcal{M}) \subseteq \mathcal{M}$ and $T(\mathcal{M}^{\perp}) \subseteq \mathcal{M}^{\perp}$.

Theorem 3.1.1 (Halmos Decomposition). *Let T be an isometry in $\mathcal{L}(\mathcal{H})$.*

1. There is a unique decomposition

$$\mathcal{H} = \mathcal{S} \oplus \mathcal{U},$$

where $\mathcal{S}$ and $\mathcal{U}$ are orthogonal subspaces that reduce T, with $T|_{\mathcal{S}}$ a shift operator on $\mathcal{S}$ and $T|_{\mathcal{U}}$ a unitary operator on $\mathcal{U}$.

2. *Define* $R^{\perp} = \mathcal{H} \ominus T\mathcal{H}$. *Then* $\{T^n(R^{\perp})\}_{n=0}^{\infty}$ *is an orthogonal sequence of subspaces of* $\mathcal{H}$, *with*

$$S = \sum_{n=0}^{\infty} \oplus T^n(R^{\perp}),$$

and

$$\mathcal{U} = \bigcap_{n=0}^{\infty} T^n(\mathcal{H}).$$

We will call $T|_S$ and $T|_{\mathcal{U}}$ the **shift** and **unitary** parts of T, respectively.

Before we prove the Halmos Decomposition, we prove a couple of useful lemmas. The following theorems are well known [7] and will be used as tools to prove the following lemma. We include them here for completeness.

Theorem 3.1.2. *A necessary and sufficient condition that the difference* $P = P_1 - P_2$ *of two projections* P_1 *and* P_2 *be a projection is that* $P_2 \le P_1$. *If this condition is satisfied and if the ranges of* P, P_1 *and* P_2 *are* $\mathcal{M}$, $\mathcal{M}_1$ *and* $\mathcal{M}_2$ *respectively, then* $\mathcal{M} = \mathcal{M}_1 \ominus \mathcal{M}_2$. *Note:* $P_2 \le P_1$ *is equivalent to* $\mathcal{M}_2 \subseteq \mathcal{M}_1$.

Theorem 3.1.3. *If* P *is an operator and if* $\{P_j\}$ *is a family of projections such that* $\sum_j P_j = P$, *then a necessary and sufficient condition that* P *be a projection is that* $P_j \perp P_k$ *whenever* $j \ne k$. *If this condition is satisfied and if, for each* j, *the range of* P_j *is the subspace* $\mathcal{M}_j$, *then the range of* P *is* $\sum_j \oplus \mathcal{M}_j$.

Theorem 3.1.4. *Let* $(P_n)_{n=1}^{\infty}$ *be a monotone decreasing sequence of projections defined on a Hilbert space* $\mathcal{H}$. *Then the sequence* $(P_n)_{n=1}^{\infty}$ *converges strongly to an operator* P, *which is a projection defined on* $\mathcal{H}$ *with range* $\bigcap_{n=1}^{\infty} P_n(\mathcal{H})$.

Lemma 3.1.1. *Suppose that* T *is an isometry in* $\mathcal{L}(\mathcal{H})$. *Then,*

$$\mathcal{H} = \left(\sum_{n=0}^{\infty} \oplus \left(T^n(\mathcal{H}) \ominus T^{n+1}(\mathcal{H}) \right) \right) \oplus \left(\bigcap_{n=0}^{\infty} T^n(\mathcal{H}) \right).$$

Proof. Henceforth, P_M will denote the projection from $\mathcal{H}$ onto the (closed) subspace M.

$$
\begin{aligned}
P_{\mathcal{H}} &= (P_{\mathcal{H}} - P_{T(\mathcal{H})}) + P_{T(\mathcal{H})} \\
&= P_{\mathcal{H} \ominus T(\mathcal{H})} + P_{T(\mathcal{H})} \\
&= P_{\mathcal{H} \ominus T(\mathcal{H})} + (P_{T(\mathcal{H})} - P_{T^2(\mathcal{H})}) + P_{T^2(\mathcal{H})} \\
&= P_{\mathcal{H} \ominus T(\mathcal{H})} + P_{T(\mathcal{H}) \ominus T^2(\mathcal{H})} + P_{T^2(\mathcal{H})} \\
&= \cdots \\
&= \sum_{k=0}^{l} P_{T^k(\mathcal{H}) \ominus T^{k+1}(\mathcal{H})} + P_{T^{l+1}(\mathcal{H})}.
\end{aligned}
$$

Letting l go to infinity, we get

$$P_{\mathcal{H}} \quad = \quad \sum_{k=0}^{\infty} P_{T^k(\mathcal{H}) \ominus T^{k+1}(\mathcal{H})} + P_{\bigcap_{n=0}^{\infty} T^n(\mathcal{H})}. \qquad (3.1)$$

It then follows that

$$\mathcal{H} \quad = \quad \left(\sum_{n=0}^{\infty} \oplus \left(T^n(\mathcal{H}) \ominus T^{n+1}(\mathcal{H}) \right) \right) \oplus \left(\bigcap_{n=0}^{\infty} T^n(\mathcal{H}) \right), \qquad (3.2)$$

as desired. $\qquad \square$

Lemma 3.1.2. *Suppose that T is an isometry in $\mathcal{L}(\mathcal{H})$. Then, $T^n \left(\mathcal{H} \ominus T(\mathcal{H}) \right) = T^n(\mathcal{H}) \ominus T^{n+1}(\mathcal{H})$.*

Proof. ($\subseteq$) Let $f \in T^n \left(\mathcal{H} \ominus T(\mathcal{H}) \right)$. Then, $f \in T^n(\mathcal{H})$ and $f = T^n f_n$ for some $f_n \in \mathcal{H} \ominus T(\mathcal{H})$. To see that $f \perp T^{n+1}(\mathcal{H})$, let $g \in T^{n+1}(\mathcal{H})$. Then, $g = T^{n+1} g_{n+1}$ for some $g_{n+1} \in \mathcal{H}$. Now, consider

$$(f, g)_{\mathcal{H}} = \left(T^n f_n, T^{n+1} g_{n+1} \right)_{\mathcal{H}} = (f_n, T g_{n+1})_{\mathcal{H}} = 0.$$

Therefore, $f \in T^n(\mathcal{H}) \ominus T^{n+1}(\mathcal{H})$. Hence, $T^n \left(\mathcal{H} \ominus T(\mathcal{H}) \right) \subseteq T^n(\mathcal{H}) \ominus T^{n+1}(\mathcal{H})$.

($\supseteq$) Let $f \in T^n(\mathcal{H}) \ominus T^{n+1}(\mathcal{H})$. Then, $f = T^n f_n$ for some $f_n \in \mathcal{H}$ and $f \perp T^{n+1}(\mathcal{H})$. To see that $f_n \perp T(\mathcal{H})$, let $g \in T(\mathcal{H})$. Then, $g = Th$ for some $h \in \mathcal{H}$. Now, consider

$$(f_n, g)_{\mathcal{H}} = (f_n, Th)_{\mathcal{H}} = \left(T^n f_n, T^{n+1} h \right)_{\mathcal{H}} = \left(f, T^{n+1} h \right)_{\mathcal{H}} = 0.$$

Therefore, $f_n \in \mathcal{H} \ominus T(\mathcal{H})$ and $f \in T^n \left(\mathcal{H} \ominus T(\mathcal{H}) \right)$. Hence, $T^n \left(\mathcal{H} \ominus T(\mathcal{H}) \right) \supseteq T^n(\mathcal{H}) \ominus T^{n+1}(\mathcal{H})$. $\qquad \square$

We are now ready to prove the Halmos Decomposition.

Proof. By the lemmas above, we have that

$$\mathcal{H} = \left(\sum_{n=0}^{\infty} \oplus T^n \left(R^{\perp} \right) \right) \oplus \left(\bigcap_{n=0}^{\infty} T^n(\mathcal{H}) \right).$$

Now, let

$$\mathcal{S} = \sum_{n=0}^{\infty} \oplus T^n \left(R^{\perp} \right).$$

By the definition of $\mathcal{S}$, $T(\mathcal{S}) \subseteq \mathcal{S}$. Next, we will show that $T|_{\mathcal{S}}$ is a shift operator on $\mathcal{S}$. To see this, let $f \in \mathcal{S}$. Then, $f = \sum_{n=0}^{\infty} T^n (f_n)$, where $f_n \in R^{\perp}$ for all n in $\mathbb{N}$. Since $(T^n(f_n))_{n=0}^{\infty}$ is an orthogonal sequence, we know that this sum converges if and only if the sum of the numeric sequence $\left(\|T^n(f_n)\|_{\mathcal{H}}^2 \right)_{n=0}^{\infty}$

converges. Since T is an isometry, that is equivalent to the sum of the numeric sequence $\left(\|f_n\|_{\mathcal{H}}^2\right)_{n=0}^{\infty}$ converging. Now, consider

$$\|T^{*n}f\|_{\mathcal{H}}^2 = \left\|\sum_{k=0}^{\infty} T^{*n}T^n(f_k)\right\|_{\mathcal{H}}^2 = \left\|\sum_{k=n}^{\infty} T^{k-n}(f_k)\right\|_{\mathcal{H}}^2 = \sum_{k=n}^{\infty}\left\|T^{k-n}(f_k)\right\|_{\mathcal{H}}^2$$

$$= \sum_{k=n}^{\infty}\|f_k\|_{\mathcal{H}}^2.$$

Since, $\sum_{k=0}^{\infty}\|f_k\|_{\mathcal{H}}^2$ converges, if follows that $\|T^{*n}f\|_{\mathcal{H}} \to 0$ as $n \to \infty$. Therefore, $T|_{\mathcal{S}}$ is a shift operator on $\mathcal{S}$, as desired.

Now, let

$$\mathcal{U} = \bigcap_{n=0}^{\infty} T^n(\mathcal{H}).$$

Then, by the lemma above, $\mathcal{S}$ and $\mathcal{U}$ are orthogonal subspaces. Next, we will show that $T(\mathcal{U}) \subseteq \mathcal{U}$. Let $g \in \mathcal{U}$. Then, $g = T^n g_n$, with $g_n \in \mathcal{H}$ for all $n \in \mathbb{N}$. Therefore, $Tg = T^{n+1}g_n$ for all $n \in \mathbb{N}$. Therefore, $Tg \in \mathcal{U}$. Hence, $T(\mathcal{U}) \subseteq \mathcal{U}$. Consequently, $\mathcal{S}$ and $\mathcal{U}$ are orthogonal subspaces that reduce T. Next, we show that $T|_{\mathcal{U}}$ is a unitary operator on $\mathcal{U}$. To see this, let $g \in \mathcal{U}$. We will now produce an h in $\mathcal{U}$ such that $Th = g$. Let $h = T^*g$. Then, by definition of T^*, $h \in \mathcal{H}$. Since $g \in \mathcal{U}$, for each n in $\mathbb{N}$, there exits g_n in $\mathcal{H}$ such that $T^n g_n = g$ for all $n \in \mathbb{N}$. Therefore, $h = T^*g = T^*T^{n+1}g_{n+1} = T^n g_{n+1}$ for all n in $\mathbb{N}$. Hence, $h \in T^n(\mathcal{H})$ for all $n \in \mathbb{N}$. Therefore, $h \in \mathcal{U}$. So, $T|_{\mathcal{U}}$ is a unitary operator on $\mathcal{U}$.

It remains to show that this decomposition is unique. To see this, suppose that $\mathcal{H} = \mathcal{S} \oplus \mathcal{U}$ and $\mathcal{H} = \mathcal{S}_1 \oplus \mathcal{U}_1$, where $\mathcal{S}, \mathcal{U}, \mathcal{S}_1$ and $\mathcal{U}_1$ satisfy the conditions of our theorem. Observe that if $f \in \mathcal{H}$ and $\|T^{*n}f\|_{\mathcal{H}} \to 0$, as $n \to \infty$, then $f \in \mathcal{S} \cap \mathcal{S}_1$. Indeed, since $f \in \mathcal{H} = \mathcal{S} \oplus \mathcal{U}$, it follow that f has the unique decomposition $f = f_{\mathcal{S}} + f_{\mathcal{U}}$, where $f_{\mathcal{S}} \in \mathcal{S}$ and $f_{\mathcal{U}} \in \mathcal{U}$. Now consider

$$\|T^{*n}f\|_{\mathcal{H}} = \|T^{*n}(f_{\mathcal{S}} + f_{\mathcal{U}})\|_{\mathcal{H}} \geq \|T^{*n}f_{\mathcal{U}}\|_{\mathcal{H}} - \|T^{*n}f_{\mathcal{S}}\|_{\mathcal{H}}.$$

Therefore,

$$\|T^{*n}f_{\mathcal{U}}\|_{\mathcal{H}} \leq \|T^{*n}f\|_{\mathcal{H}} + \|T^{*n}f_{\mathcal{S}}\|_{\mathcal{H}}.$$

Now, letting $n \to \infty$, we see that $\|T^{*n}f_{\mathcal{U}}\|_{\mathcal{H}} \to 0$, as $n \to \infty$, but since $f_{\mathcal{U}} \in \mathcal{U}$, it follows that $\|T^{*n}f_{\mathcal{U}}\|_{\mathcal{H}} = \|f_{\mathcal{U}}\|_{\mathcal{H}}$ for all $n \in \mathbb{N}$. Therefore, $\|f_{\mathcal{U}}\|_{\mathcal{H}} = 0$ and so $f_{\mathcal{U}} = 0$. Hence, $f = f_{\mathcal{S}}$ as desired. An analogous argument would show that $f \in \mathcal{S}_1$. Now that this has been established, it follows that $\mathcal{S} = \mathcal{S}_1$, since if $f \in \mathcal{S}$ then $\|T^{*n}f\|_{\mathcal{H}} \to 0$, as $n \to \infty$. Therefore, $f \in \mathcal{S}_1$ and so $\mathcal{S} \subseteq \mathcal{S}_1$. Similarly, $\mathcal{S}_1 \subseteq \mathcal{S}$. Therefore, $\mathcal{S} = \mathcal{S}_1$ and hence, $\mathcal{U} = \mathcal{U}_1$. It then follows that this decomposition is unique. $\square$

3.2   Invariant Subspaces of $L^2(\mathbb{T})$

We will now use the Halmos Decomposition to identify all subspaces of $L^2(\mathbb{T})$ that are invariant under multiplication by the coordinate function. We start by recalling some notations and terminology.

Let $\mathbb{T} = \{e^{i\lambda} : \lambda \in [-\pi, \pi)\}$. We will write $L^2(\mathbb{T})$ to denote the collection of all Lebesgue measurable functions $f : \mathbb{T} \to \mathbb{C}$ such that

$$\int_{[-\pi,\pi)} |f(e^{i\lambda})|^2 \, d\sigma(e^{i\lambda}) < \infty.$$

Here, like before, σ denotes normalized Lebesgue measure on $\mathbb{T}$. As always, we will identify functions that are equal $[\sigma]$-a.e. Define

$$\hat{f}(n) = (f, e_n)_{L^2(\mathbb{T})} = \int_{[-\pi,\pi)} f(e^{i\lambda}) e^{-in\lambda} \, d\sigma(e^{i\lambda}),$$

for all $n \in \mathbb{Z}$, where $e_n(e^{i\lambda}) = e^{in\lambda}$. These are the Fourier coefficients of f. An important subspace of $L^2(\mathbb{T})$ that will come into play is

$$H^2(\mathbb{T}) = \left\{ f \in L^2(\mathbb{T}) : \hat{f}(n) = 0, \text{ for } n = -1, -2, \dots \right\}.$$

Let $S : L^2(\mathbb{T}) \to L^2(\mathbb{T})$ be defined by $S(f)(e^{i\lambda}) = e^{i\lambda} f(e^{i\lambda})$. It is straightforward to verify that S is an isometry in $\mathcal{L}(L^2(\mathbb{T}))$. A subspace $\mathcal{M}$ of $L^2(\mathbb{T})$ is said to be S-invariant if $S(\mathcal{M}) \subseteq \mathcal{M}$. If $S|_{\mathcal{M}}$ is unitary on $\mathcal{M}$, then Wiener showed the following.

Theorem 3.2.1. *Suppose that $\mathcal{M}$ is an S-invariant subspace of $L^2(\mathbb{T})$. Then, $S(\mathcal{M}) = \mathcal{M}$ if and only if $\mathcal{M} = 1_E L^2(\mathbb{T})$, where E is a measurable subset of $\mathbb{T}$ and 1_E is the indicator function of E.*

We will call such S-invariant subspaces doubly invariant subspaces.

Proof: We will leave it to the reader to verify that every subspace of the form $1_E L^2(\mathbb{T})$, where E is a measurable subset of $\mathbb{T}$, is a doubly invariant subspace of $L^2(\mathbb{T})$. We will prove the other direction.

Let 1 denote the function in $L^2(\mathbb{T})$ that takes the value 1 at every point of $\mathbb{T}$. Let $q = P_{\mathcal{M}} 1$, the projection of 1 onto $\mathcal{M}$. Then, $1 - q$ is orthogonal to $\mathcal{M}$ and hence orthogonal to $S^n q$ for all n in $\mathbb{Z}$. It follows from this that $q - |q|^2 = 0$ $[\sigma]$-a.e. Therefore, q takes values 0 and 1 $[\sigma]$-a.e. Let E denote the subset of $\mathbb{T}$ where q takes the value 1. This set is Lebesgue measurable and $q = 1_E$ $[\sigma]$-a.e. We will now show that $\mathcal{M} = qL^2(\mathbb{T})$. Since q is in $\mathcal{M}$ and $\mathcal{M}$ is doubly invariant, it follows that $qL^2(\mathbb{T}) \subseteq \mathcal{M}$. If this inclusion is strict, then one can find a function in $\mathcal{M} \ominus qL^2(\mathbb{T})$ that is not identically zero. Let us call this function g. Therefore, g is orthogonal to $S^n q$ for all $n \in \mathbb{Z}$ and hence, $\bar{g}q = 0$ $[\sigma]$-a.e. We also observe that since $\mathbf{1} - q$ is orthogonal to $\mathcal{M}$ and $g \in \mathcal{M}$ then $\mathbf{1} - q$ is orthogonal to $S^n g$ for all $n \in \mathbb{Z}$ and hence $(\mathbf{1} - q)\bar{g} = 0$ $[\sigma]$-a.e. Therefore, $g = \bar{g}q + (\mathbf{1} - q)\bar{g} = 0$ $[\sigma]$-a.e. Therefore, the containment must not be strict and $\mathcal{M} = qL^2(\mathbb{T})$ as desired. $\square$

Now, suppose that $S(\mathcal{M}) \subsetneq \mathcal{M}$. Then, $R^{\perp} := \mathcal{M} \ominus S(\mathcal{M}) \neq \{0\}$. Let φ be a nonzero member of $R^{\perp}$. By the Halmos Decomposition, it follows that $(S^n \varphi, \varphi)_{L^2(\mathbb{T})} = 0$ and $(\varphi, S^n \varphi)_{L^2(\mathbb{T})} = 0$ for $n > 0$. These observations show that $|\varphi|^2$ is constant $[\sigma]$-a.e. and since φ is nonzero, this constant value must also be nonzero. We will now show that the dimension of $R^{\perp}$ is one. Since $R^{\perp} \neq \{0\}$, we know that the dimension of $R^{\perp}$ is at least one. Suppose that it is more than one. Then, there exists $\varphi \neq 0$ and $\psi \neq 0$ in $R^{\perp}$ with $\varphi \perp \psi$. Our previous calculation also shows us that neither function can vanish on a set of positive Lebesgue measure. Now, by the Halmos Decomposition and the fact that $\varphi \perp \psi$, it follows that $(S^n \varphi, \psi)_{L^2(\mathbb{T})} = 0$ and $(\varphi, S^n \psi)_{L^2(\mathbb{T})} = 0$ for $n \geq 0$. Therefore, we may conclude that $\varphi\bar{\psi} = 0$, but this is a contradiction, since neither function can vanish on a set of positive Lebesgue measure. Therefore, our supposition must be incorrect. Hence, the dimension of $R^{\perp}$ is one, as desired. If we now choose φ in $R^{\perp}$ with $\|\varphi\|_{L^2(\mathbb{T})} = 1$, then $|\varphi| = 1$ $[\sigma]$-a.e. Functions that have modulus one $[\sigma]$-a.e. are called unimodular functions. So, based on our work above, we have that $R^{\perp} = \{\alpha\varphi : \alpha \in \mathbb{C}\}$. Therefore,

$$\sum_{n=0}^{\infty} \oplus S^n(R^{\perp}) = \sum_{n=0}^{\infty} \oplus S^n(\{\alpha\varphi : \alpha \in \mathbb{C}\})$$
$$= \varphi H^2(\mathbb{T}).$$

Next, we will show that the unitary part of this Halmos Decomposition is zero. To see this, consider the doubly invariant subspace $\mathcal{H}$ spanned by the orthonormal sequence $(e_n \varphi)_{n \in \mathbb{Z}}$. By Wiener's Theorem and the fact that φ does not vanish on a set of positive Lebesgue measure gives us that $\mathcal{H} = L^2(\mathbb{T})$. Now, $(e_n \varphi)_{n=0}^{\infty}$ spans $\varphi H^2(\mathbb{T})$ and $(e_n \varphi)_{n=-\infty}^{-1}$ is orthogonal to $\mathcal{M}$ since that is equivalent to φ being orthogonal to $S^n \mathcal{M}$ for $n = 1, 2, \ldots$, which follows from the definition of φ. It follows then that $\mathcal{M} = \varphi H^2(\mathbb{T})$ and therefore, the unitary part of the Halmos Decomposition must be zero. This work proves one direction of the following theorem. We leave it to the reader to prove the other direction.

Theorem 3.2.2. *Suppose that $\mathcal{M}$ is an S-invariant subspace of $L^2(\mathbb{T})$. Then, $S(\mathcal{M}) \subsetneq \mathcal{M}$ if and only if $\mathcal{M} = \varphi H^2(\mathbb{T})$, where φ is a unimodular function.*

We will call such S-invariant subspaces simply invariant subspaces. This theorem was first proved by Helson and Lowdenslager.

3.3 Invariant Subspaces of $H^2(\mathbb{T})$

The following theorem is a corollary of the last theorem in the previous section.

Corollary 3.3.1. $\mathcal{M} \neq \{0\}$ *is an S-invariant subspace of $H^2(\mathbb{T})$ if and only if $\mathcal{M} = \varphi H^2(\mathbb{T})$, where φ is an inner function.*

By an inner function, we mean a function in $H^2(\mathbb{T})$ that has modulus one $[\sigma]$-a.e.

One can see this, once we make a couple of observations. First, we show that the unitary part of any subspace of $H^2(\mathbb{T})$ is always zero. To see this, suppose that $f \in \bigcap_{n=0}^{\infty} S^n(\mathcal{M})$, where $\mathcal{M}$ is the subspace in question. Then, $f = S^n f_n$ for every $n \in \mathbb{N}$, where $f_n \in \mathcal{M} \subset H^2(\mathbb{T})$. It follows from this that $\hat{f}(n) = 0$ for all $n \in \mathbb{N}$. Therefore, $f = 0$. Now that this is established, it follows from the above theorem that $\mathcal{M} = \varphi H^2(\mathbb{T})$, where φ is a unimodular function. Now, since φ is in $H^2(\mathbb{T})$, it follows that φ is an inner function. This corollary is typically called Beurling's Theorem. He was the first to prove this theorem, but used a different approach to prove it. The $L^2(\mathbb{T})$ result was not known at the time Beurling gave his proof.

3.4 The Halmos Fourfold Decomposition

In this section, we will generalize the Halmos Decomposition for a pair of isometries. We begin by recalling the statement of the Halmos Decomposition.

Theorem 3.4.1 (Halmos Decomposition). *Let T be an isometry in $\mathcal{L}(\mathcal{H})$.*

1. *There is a unique decomposition*

$$\mathcal{H} = \mathcal{S} \oplus \mathcal{U},$$

   where $\mathcal{S}$ and $\mathcal{U}$ are orthogonal subspaces that reduce T, with $T|_{\mathcal{S}}$ a shift operator on $\mathcal{S}$ and $T|_{\mathcal{U}}$ a unitary operator on $\mathcal{U}$.

2. *Define $R^{\perp} = \mathcal{H} \ominus T\mathcal{H}$. Then $\{T^n(R^{\perp})\}_{n=0}^{\infty}$ is an orthogonal sequence of subspaces of $\mathcal{H}$, with*

$$\mathcal{S} = \sum_{n=0}^{\infty} \oplus T^n(R^{\perp}),$$

   and

$$\mathcal{U} = \bigcap_{n=0}^{\infty} T^n(\mathcal{H}).$$

We will call $T|_{\mathcal{S}}$ and $T|_{\mathcal{U}}$ the shift and unitary parts of T, respectively.

Now, let T_1 and T_2 be a pair of isometries in $\mathcal{L}(\mathcal{H})$. As such, each has a Halmos Decomposition. For T_1, we have

$$\mathcal{H} = \mathcal{S}_1 \oplus \mathcal{U}_1,$$

where $\mathcal{S}_1$ and $\mathcal{U}_1$ are orthogonal subspaces that reduce T_1, with $T_1|_{\mathcal{S}_1}$ a shift operator on $\mathcal{S}_1$ and $T_1|_{\mathcal{U}_1}$ a unitary operator on $\mathcal{U}_1$ and for T_2, we have

$$\mathcal{H} = \mathcal{S}_2 \oplus \mathcal{U}_2,$$

where $\mathcal{S}_2$ and $\mathcal{U}_2$ are orthogonal subspaces that reduce T_2, with $T_2|_{\mathcal{S}_2}$ a shift operator on $\mathcal{S}_2$ and $T_2|_{\mathcal{U}_2}$ a unitary operator on $\mathcal{U}_2$.

Our next goal is to try to combine these decompositions into a single decomposition. We will use the following lemma to help us accomplish our goal.

Lemma 3.4.1. *Let $A_1 = \{P_{\mathcal{S}_1}, P_{\mathcal{U}_1}\}$ and $A_2 = \{P_{\mathcal{S}_2}, P_{\mathcal{U}_2}\}$. If any projection in set A_1 commutes with any projection in set A_2, then every projection in set A_1 commutes with every projection in set A_2.*

This lemma follows from a well-known theorem found in [7].

Theorem 3.4.2. *If P is a projection on a subspace $\mathcal{M}$, then $1 - P$ is a projection on $\mathcal{M}^{\perp}$.*

We will leave it to the reader to use this theorem to prove our lemma. We are now ready to state this new combined decomposition.

Theorem 3.4.3. *Let T_1 and T_2 be as stated above and let the hypothesis of Lemma 3.4.1 be satisfied. Then, $\mathcal{H}$ has the following decomposition.*

$$\mathcal{H} = (\mathcal{S}_1 \cap \mathcal{S}_2) \oplus (\mathcal{S}_1 \cap \mathcal{U}_2) \oplus (\mathcal{S}_2 \cap \mathcal{U}_1) \oplus (\mathcal{U}_1 \cap \mathcal{U}_2).$$

The proof of this theorem follows from another well-known theorem, again found in [7].

Theorem 3.4.4. *A necessary and sufficient condition that the product $P = P_1 P_2$ of two projections P_1 and P_2 be a projection is that P_1 commutes with P_2. If this condition is satisfied and if the ranges of P, P_1 and P_2 are $\mathcal{M}$, $\mathcal{M}_1$ and $\mathcal{M}_2$ respectively, then $\mathcal{M} = \mathcal{M}_1 \cap \mathcal{M}_2$.*

At this point, a few questions might come to mind. How do T_1 and T_2 behave on this decomposition? Do these subspaces reduce T_1 and T_2? Are T_1 and T_2 shifts on certain subspaces and unitary on others? We will start by examining when these subspaces reduce T_1 and T_2. The following theorem is well known, see [7].

Theorem 3.4.5. *A subspace* $\mathcal{M}$, *with projection* P, *reduces an operator* A *if and only if* P *commutes with* A.

Theorem 3.4.6. *The subspaces given in Theorem 3.4.3, regarding the decomposition of* $\mathcal{H}$, *reduce* T_1 *and* T_2 *if and only if* T_1 *and* T_2 *commute with every projection in the set* $\{P_{\mathcal{S}_1 \cap \mathcal{S}_2}, P_{\mathcal{S}_1 \cap \mathcal{U}_2}, P_{\mathcal{U}_1 \cap \mathcal{S}_2}, P_{\mathcal{U}_1 \cap \mathcal{U}_2}\}$.

Using the sets defined in Lemma 3.4.1, we have the following theorem.

Theorem 3.4.7. *The subspaces given in Theorem 3.4.3, regarding the decomposition of* $\mathcal{H}$, *reduce* T_1 *and* T_2 *if* T_1 *commutes with either projection in set* A_2 *(hence both) and* T_2 *commutes with either projection in set* A_1 *(hence both).*

Under any conditions for which the subspaces of Theorem 3.4.3 reduce T_1 and T_2, it is straightforward to verify that

1. $T_1|_{\mathcal{S}_1 \cap \mathcal{S}_2}$ is a shift operator on $\mathcal{S}_1 \cap \mathcal{S}_2$.

2. $T_2|_{\mathcal{S}_1 \cap \mathcal{S}_2}$ is a shift operator on $\mathcal{S}_1 \cap \mathcal{S}_2$.

3. $T_1|_{\mathcal{S}_1 \cap \mathcal{U}_2}$ is a shift operator on $\mathcal{S}_1 \cap \mathcal{U}_2$.

4. $T_2|_{\mathcal{S}_1 \cap \mathcal{U}_2}$ is a unitary operator on $\mathcal{S}_1 \cap \mathcal{U}_2$.

5. $T_1|_{\mathcal{S}_2 \cap \mathcal{U}_1}$ is a unitary operator on $\mathcal{S}_2 \cap \mathcal{U}_1$.

6. $T_2|_{\mathcal{S}_2 \cap \mathcal{U}_1}$ is a shift operator on $\mathcal{S}_2 \cap \mathcal{U}_1$.

7. $T_1|_{\mathcal{U}_1 \cap \mathcal{U}_2}$ is a unitary operator on $\mathcal{U}_1 \cap \mathcal{U}_2$.

8. $T_2|_{\mathcal{U}_1 \cap \mathcal{U}_2}$ is a unitary operator on $\mathcal{U}_1 \cap \mathcal{U}_2$.

For this reason, we will henceforth write $\mathcal{H}_{ss}$ for $\mathcal{S}_1 \cap \mathcal{S}_2$, $\mathcal{H}_{su}$ for $\mathcal{S}_1 \cap \mathcal{U}_2$, $\mathcal{H}_{us}$ for $\mathcal{U}_1 \cap \mathcal{S}_2$, and $\mathcal{H}_{uu}$ for $\mathcal{U}_1 \cap \mathcal{U}_2$. So, our decomposition in Theorem 3.4.3, in the case where the subspaces of Theorem 3.4.3 reduce T_1 and T_2, will be written as

$$\mathcal{H} = \mathcal{H}_{ss} \oplus \mathcal{H}_{su} \oplus \mathcal{H}_{us} \oplus \mathcal{H}_{uu}.$$

We will call this decomposition the Halmos Fourfold Decomposition. Our next job will be to show that if $\mathcal{H}$ has a Halmos Fourfold Decomposition, then the decomposition is unique.

Proposition 3.4.1. *If a Halmos Fourfold Decomposition exists for a pair of isometries, then the decomposition is unique.*

Proof: Suppose that

$$\mathcal{H} = \mathcal{H}_{ss} \oplus \mathcal{H}_{su} \oplus \mathcal{H}_{us} \oplus \mathcal{H}_{uu}$$

and

$$\mathcal{H} = \mathcal{H}'_{ss} \oplus \mathcal{H}'_{su} \oplus \mathcal{H}'_{us} \oplus \mathcal{H}'_{uu}$$

are both Halmos Fourfold Decompositions for a pair of commuting isometries. Observe that

$$\mathcal{H} = [\mathcal{H}_{ss} \oplus \mathcal{H}_{su}] \oplus [\mathcal{H}_{us} \oplus \mathcal{H}_{uu}]$$

is the Halmos Decomposition for T_1 and

$$\mathcal{H} = [\mathcal{H}'_{ss} \oplus \mathcal{H}'_{su}] \oplus [\mathcal{H}'_{us} \oplus \mathcal{H}'_{uu}]$$

is the Halmos Decomposition for T_1. Similarly,

$$\mathcal{H} = [\mathcal{H}_{ss} \oplus \mathcal{H}_{us}] \oplus [\mathcal{H}_{su} \oplus \mathcal{H}_{uu}]$$

is the Halmos Decomposition of T_2 and

$$\mathcal{H} = [\mathcal{H}'_{ss} \oplus \mathcal{H}'_{us}] \oplus [\mathcal{H}'_{su} \oplus \mathcal{H}'_{uu}]$$

is the Halmos Decomposition of T_2. Now, by the uniqueness of the Halmos Decomposition, we get that

$$\mathcal{H}_{ss} \oplus \mathcal{H}_{su} = \mathcal{H}'_{ss} \oplus \mathcal{H}'_{su} \perp \mathcal{H}'_{us} \oplus \mathcal{H}'_{uu} = \mathcal{H}_{us} \oplus \mathcal{H}_{uu}$$

and

$$\mathcal{H}_{ss} \oplus \mathcal{H}_{us} = \mathcal{H}'_{ss} \oplus \mathcal{H}'_{us} \perp \mathcal{H}'_{su} \oplus \mathcal{H}'_{uu} = \mathcal{H}_{su} \oplus \mathcal{H}_{uu}.$$

It follows from these equations that $\mathcal{H}_{ss} \perp \mathcal{H}'_{us}$ and $\mathcal{H}_{ss} \subseteq \mathcal{H}'_{ss} \oplus \mathcal{H}'_{us}$. Therefore, $\mathcal{H}_{ss} \subseteq \mathcal{H}'_{ss}$. Also, we see that $\mathcal{H}'_{ss} \perp \mathcal{H}_{us}$ and $\mathcal{H}'_{ss} \subseteq \mathcal{H}_{ss} \oplus \mathcal{H}_{us}$. Therefore, $\mathcal{H}'_{ss} \subseteq \mathcal{H}_{ss}$. Consequently, $\mathcal{H}'_{ss} = \mathcal{H}_{ss}$. In an analogous way, one sees that $\mathcal{H}'_{su} = \mathcal{H}_{su}$, $\mathcal{H}'_{us} = \mathcal{H}_{us}$, and $\mathcal{H}'_{uu} = \mathcal{H}_{uu}$. Therefore, our decomposition is unique as desired. □

Our next goal will be to determine the existence of the Halmos Fourfold Decomposition in the context of the behavior of the isometries on $\mathcal{H}$.

We say V_1 and V_2 in $\mathcal{L}(\mathcal{H})$ doubly commute if V_1 commutes with V_2 and V_1^* commutes with V_2. Note that V_1^* commutes with V_2 if and only if V_1 commutes with V_2^*.

Theorem 3.4.8. *Every pair of doubly commuting isometries has a Halmos Fourfold Decomposition.*

Proof: Let T_1 and T_2 be our pair of doubly commuting isometries. Let $\mathcal{T} = \{T_1, T_1^*\}$ and let $\mathcal{T}'$ denote the commutant of $\mathcal{T}$. Since T_1 and T_2 doubly commute, it follows that $T_2^n T_2^{*n}$ is in $\mathcal{T}'$ for all $n \in \mathbb{N}$. It is well known that the commutant of a set of operators is strongly closed. Therefore, the strong limit of $T_2^n T_2^{*n}$, which is $P_{\mathcal{U}_2}$, is in $\mathcal{T}'$. Therefore, T_1 commutes with $P_{\mathcal{U}_2}$. An analogous argument shows that T_2 commutes with $P_{\mathcal{U}_1}$. These observations are part of the ingredients necessary to employ Theorem 3.4.7. It remains to show that $P_{\mathcal{U}_1}$ commutes with $P_{\mathcal{U}_2}$. Once this is done, we have the existence of the Halmos Fourfold Decomposition, as desired. To see that $P_{\mathcal{U}_1}$ commutes

with $P_{\mathcal{U}_2}$ we only need to observe that $\mathcal{T}''$, the double commutant of $\mathcal{T}$, contains $P_{\mathcal{U}_1}$ and that $\mathcal{T}' = \mathcal{T}'''$. So therefore, $P_{\mathcal{U}_1}$ commutes with $P_{\mathcal{U}_2}$, as desired.
□

Our next goal will be to examine each subspace in the Halmos Fourfold Decomposition and find simple and useful ways to represent each of them. We will study them under the assumption that T_1 and T_2 doubly commute.

Based on our work above and the Halmos Decomposition, we certainly have the following descriptions.

$$\mathcal{H}_{ss} = \left[\sum_{n=0}^{\infty} \oplus T_1^n\left(R_1^{\perp}\right)\right] \cap \left[\sum_{n=0}^{\infty} \oplus T_2^n\left(R_2^{\perp}\right)\right]$$

$$\mathcal{H}_{su} = \left[\sum_{n=0}^{\infty} \oplus T_1^n\left(R_1^{\perp}\right)\right] \cap \left[\bigcap_{n=0}^{\infty} T_2^n(\mathcal{H})\right]$$

$$\mathcal{H}_{us} = \left[\bigcap_{n=0}^{\infty} T_1^n(\mathcal{H})\right] \cap \left[\sum_{n=0}^{\infty} \oplus T_2^n\left(R_2^{\perp}\right)\right]$$

$$\mathcal{H}_{uu} = \left[\bigcap_{n=0}^{\infty} T_1^n(\mathcal{H})\right] \cap \left[\bigcap_{n=0}^{\infty} T_2^n(\mathcal{H})\right],$$

where $R_i^{\perp} = \mathcal{H} \ominus T_i(\mathcal{H})$, for $i = 1, 2$.

We will show that under the doubly commuting condition the following descriptions are also accurate.

$$\mathcal{H}_{ss} = \sum_{n=0}^{\infty}\sum_{m=0}^{\infty} \oplus T_1^n T_2^m\left(R_1^{\perp} \cap R_2^{\perp}\right)$$

$$\mathcal{H}_{su} = \sum_{n=0}^{\infty} \oplus T_1^n\left(\bigcap_{m=0}^{\infty} T_2^m(R_1^{\perp})\right)$$

$$\mathcal{H}_{us} = \sum_{n=0}^{\infty} \oplus T_2^n\left(\bigcap_{m=0}^{\infty} T_1^m(R_2^{\perp})\right)$$

$$\mathcal{H}_{uu} = \bigcap_{m,n=0}^{\infty} T_1^m T_2^n(\mathcal{H}).$$

Let $\mathcal{U}_1 = \bigcap_{m=0}^{\infty} T_1^m(\mathcal{H})$, $\mathcal{U}_2 = \bigcap_{n=0}^{\infty} T_2^n(\mathcal{H})$, and $\mathcal{U} = \bigcap_{m,n=0}^{\infty} T_1^m T_2^n(\mathcal{H})$. Now, we will show that $\mathcal{H}_{uu} = \mathcal{U}$. To see this, note that under the doubly commuting condition, we get that

$$P_{\mathcal{U}_1 \cap \mathcal{U}_2} = P_{\mathcal{U}_1}P_{\mathcal{U}_2} = \lim_{n\to\infty} V_1^n V_1^{*n} \lim_{m\to\infty} V_2^m V_2^{*m} = \lim_{n,m\to\infty} V_1^n V_2^m V_2^{*m} V_1^{*n} = P_{\mathcal{U}}.$$

Therefore, $\mathcal{H}_{uu} = \bigcap_{m,n=0}^{\infty} T_1^m T_2^n(\mathcal{H})$, as desired.

Lemma 3.4.2. *Let T_1 and T_2 be doubly commuting isometries and let $R_i^\perp = \mathcal{H} \ominus T_i(\mathcal{H})$, for $i = 1, 2$. Then, $R_1^\perp$ reduces T_2 and $R_2^\perp$ reduces T_1.*

Proof: $T_2 P_{R_1^\perp} = T_2(1 - T_1 T_1^*) = (1 - T_1 T_1^*) T_2 = P_{R_1^\perp} T_2$. Therefore, by Theorem 3.4.5, $R_1^\perp$ reduces T_2. The second part is proved in an analogous way. $\square$

Lemma 3.4.3. *Let T_1 and T_2 be doubly commuting isometries and let $R_i^\perp = \mathcal{H} \ominus T_i(\mathcal{H})$, for $i = 1, 2$. Then, for each n in $\mathbb{N}$,*

$$R_1^\perp \cap T_2^n(\mathcal{H}) = T_2^n(R_1^\perp)$$

and

$$R_2^\perp \cap T_1^n(\mathcal{H}) = T_1^n(R_2^\perp).$$

Proof: We will prove the first equation; the second equation is proved in an analogous way.

($\supseteq$) Since $R_1^\perp \subseteq \mathcal{H}$, $T_2^n(R_1^\perp) \subseteq T_2^n(\mathcal{H})$ and $R_1^\perp \cap T_2^n(R_1^\perp) \subseteq R_1^\perp \cap T_2^n(\mathcal{H})$. Now, by Lemma 3.4.2, $T_2^n(R_1^\perp) \subset R_1^\perp$ and therefore, $T_2^n(R_1^\perp) \subseteq R_1^\perp \cap T_2^n(\mathcal{H})$.

($\subseteq$) $x \in R_1^\perp \cap T_2^n(\mathcal{H})$. Then, $x \perp R_1$ and there exists a $y \in \mathcal{H}$ such that $x = T_2^n y$. We will show that under these conditions $y \in R_1^\perp$. To see this, note that for all $z \in R_1$,

$$(y, z)_{\mathcal{H}} = (T_2^n y, T_2^n z)_{\mathcal{H}} = (x, T_2^n z)_{\mathcal{H}}.$$

By Lemma 3.4.2, $T_2^n z \in R_1$ for all $z \in R_1$. Therefore, $(x, T_2^n z)_{\mathcal{H}} = 0$ for all $z \in R_1$. Therefore, $y \in R_1^\perp$, as desired. Hence, $x \in T_2^n(R_1^\perp)$. $\square$

Lemma 3.4.4. *Let T_1 and T_2 be doubly commuting isometries and let $R_i^\perp = \mathcal{H} \ominus T_i(\mathcal{H})$, for $i = 1, 2$. Then,*

$$R_1^\perp = \left[\sum_{n=0}^{\infty} \oplus T_2^n(R_1^\perp \cap R_2^\perp) \right] \oplus \left[\bigcap_{n=0}^{\infty} T_2^n(R_1^\perp) \right]$$

and

$$R_2^\perp = \left[\sum_{n=0}^{\infty} \oplus T_1^n(R_1^\perp \cap R_2^\perp) \right] \oplus \left[\bigcap_{n=0}^{\infty} T_1^n(R_2^\perp) \right].$$

Proof: We will prove the first equation; the second equation is proved in an analogous way. By Lemma 3.4.2, we have that $T_2(R_1^\perp) \subseteq R_1^\perp$. Therefore, we may apply the Halmos Decomposition to $T_2|_{R_1^\perp}$. Doing this, we get

$$R_1^\perp = \left[\sum_{n=0}^{\infty} \oplus T_2^n \left(R_1^\perp \ominus T_2(R_1^\perp) \right) \right] \oplus \left[\bigcap_{n=0}^{\infty} T_2^n(R_1^\perp) \right].$$

It remains to show that $R_1^\perp \ominus T_2(R_1^\perp) = R_1^\perp \cap R_2^\perp$. To see this, note that by Lemma 3.4.3,

$$P_{T_2(R_1^\perp)} = P_{R_1^\perp \cap T_2(\mathcal{H})} = P_{R_1^\perp} P_{T_2(\mathcal{H})} = (1 - T_1 T_1^*) T_2 T_2^*.$$

Therefore,

$$P_{R_1 \ominus T_2(R_1^\perp)} = (1 - T_1 T_1^*) - (1 - T_1 T_1^*) T_2 T_2^* = (1 - T_1 T_1^*)(1 - T_2 T_2^*) = P_{R_1^\perp \cap R_2^\perp}.$$

Therefore, $R_1^\perp \ominus T_2(R_1^\perp) = R_1^\perp \cap R_2^\perp$ and so

$$R_1^\perp = \left[\sum_{n=0}^{\infty} \oplus T_2^n (R_1^\perp \cap R_2^\perp) \right] \oplus \left[\bigcap_{n=0}^{\infty} T_2^n (R_1^\perp) \right],$$

as desired. $\square$

Next, we will examine the structure of $\mathcal{H}_{su}$. We already have that

$$\mathcal{H}_{su} = \left[\sum_{n=0}^{\infty} \oplus T_1^n (R_1^\perp) \right] \cap \left[\bigcap_{n=0}^{\infty} T_2^n (\mathcal{H}) \right].$$

Now, using Lemma 3.4.4, we get that

$$\mathcal{H}_{su} = \left[\sum_{n=0}^{\infty} \oplus T_1^n \left(\left[\sum_{n=0}^{\infty} \oplus T_2^n (R_1^\perp \cap R_2^\perp) \right] \oplus \left[\bigcap_{n=0}^{\infty} T_2^n (R_1^\perp) \right] \right) \right] \cap \left[\bigcap_{n=0}^{\infty} T_2^n (\mathcal{H}) \right].$$

Simplifying, we get

$$\mathcal{H}_{su} = \left[\left[\sum_{n=0}^{\infty} \sum_{k=0}^{\infty} \oplus T_1^n T_2^k (R_1^\perp \cap R_2^\perp) \right] \oplus \left[\sum_{n=0}^{\infty} \oplus T_1^n \left(\bigcap_{k=0}^{\infty} T_2^k (R_1^\perp) \right) \right] \right] \cap \left[\bigcap_{n=0}^{\infty} T_2^n (\mathcal{H}) \right].$$

Since $R_1^\perp \cap R_2^\perp \subseteq R_2^\perp$, it follows that $T_1^n (R_1^\perp \cap R_2^\perp) \subseteq T_1^n (R_2^\perp)$. By Lemma 3.4.2, it follows that $T_1^n (R_1^\perp \cap R_2^\perp) \subseteq R_2^\perp$. Therefore, $T_1^n T_2^k (R_1^\perp \cap R_2^\perp) \subseteq T_2^k (R_2^\perp)$. So from the Halmos Decomposition of T_2, it follows that $\left[\sum_{n=0}^{\infty} \sum_{k=0}^{\infty} \oplus T_1^n T_2^k (R_1^\perp \cap R_2^\perp) \right] \perp \left[\bigcap_{n=0}^{\infty} T_2^n (\mathcal{H}) \right]$. Therefore, we can simplify our description of $\mathcal{H}_{su}$ to

$$\mathcal{H}_{su} = \left[\sum_{n=0}^{\infty} \oplus T_1^n \left(\bigcap_{k=0}^{\infty} T_2^k (R_1^\perp) \right) \right] \cap \left[\bigcap_{n=0}^{\infty} T_2^n (\mathcal{H}) \right].$$

Now, since $R_1^\perp \subseteq \mathcal{H}$, it follows that $\bigcap_{k=0}^{\infty} T_2^k (R_1^\perp) \subseteq \bigcap_{k=0}^{\infty} T_2^k (\mathcal{H})$. The doubly commuting condition implies that $\bigcap_{k=0}^{\infty} T_2^k(\mathcal{H})$ is invariant under T_1. There-

fore, $T_1^n \left(\bigcap_{k=0}^{\infty} T_2^k(R_1^\perp) \right) \subseteq \bigcap_{k=0}^{\infty} T_2^k(\mathcal{H})$, for all $n \in \mathbb{N}$. Therefore, $\mathcal{H}_{su}$ can be

simplified further to

$$\mathcal{H}_{su} = \sum_{n=0}^{\infty} \oplus T_1^n \left(\bigcap_{k=0}^{\infty} T_2^k(R_1^\perp) \right).$$

An analogous argument would show that

$$\mathcal{H}_{us} = \sum_{n=0}^{\infty} \oplus T_2^n \left(\bigcap_{m=0}^{\infty} T_1^m(R_2^\perp) \right).$$

It remains to show that $\mathcal{H}_{ss}$ has the desired description. To see this, we again start with what we know, which is that

$$\mathcal{H}_{ss} = \left[\sum_{n=0}^{\infty} \oplus T_1^n \left(R_1^\perp \right) \right] \cap \left[\sum_{n=0}^{\infty} \oplus T_2^n \left(R_2^\perp \right) \right].$$

Employing Lemma 3.4.4, we have that

$$\mathcal{H}_{ss} = \left[\sum_{n=0}^{\infty} \oplus T_1^n \left(\left[\sum_{n=0}^{\infty} \oplus T_2^n(R_1^\perp \cap R_2^\perp) \right] \oplus \left[\bigcap_{n=0}^{\infty} T_2^n(R_1^\perp) \right] \right) \right]$$

$$\cap \left[\sum_{n=0}^{\infty} \oplus T_2^n \left(\left[\sum_{n=0}^{\infty} \oplus T_1^n(R_1^\perp \cap R_2^\perp) \right] \oplus \left[\bigcap_{n=0}^{\infty} T_1^n(R_2^\perp) \right] \right) \right].$$

Simplifying, we get that

$$\mathcal{H}_{ss} = \left[\left[\sum_{n=0}^{\infty} \sum_{k=0}^{\infty} \oplus T_1^n T_2^k(R_1^\perp \cap R_2^\perp) \right] \oplus \left[\sum_{n=0}^{\infty} \oplus T_1^n \left(\bigcap_{k=0}^{\infty} T_2^k(R_1^\perp) \right) \right] \right]$$

$$\cap \left[\left[\sum_{n=0}^{\infty} \sum_{k=0}^{\infty} \oplus T_1^n T_2^k(R_1^\perp \cap R_2^\perp) \right] \oplus \left[\sum_{n=0}^{\infty} \oplus T_2^n \left(\bigcap_{k=0}^{\infty} T_1^k(R_2^\perp) \right) \right] \right].$$

Let us write the above equation as

$$\mathcal{H}_{ss} = [A \oplus B] \cap [C \oplus D].$$

Using this form of our equation, we see that $A \perp B$, $C \perp D$ and from the original equation, we see that $A = C$. Therefore, we have

$$\mathcal{H}_{ss} = [A \oplus B] \cap [A \oplus D],$$

with $A \perp B$ and $A \perp D$. If we can show that $B \perp D$, then our equation simplifies to

$$\mathcal{H}_{ss} = A,$$

which is our desired description. To see that $B \perp D$, note that if $f \in B$ and $g \in D$, then

$$f = \sum_{n=0}^{\infty} T_1^n y_n,$$

where $y_n \in \bigcap_{k=0}^{\infty} T_2^k(R_1^{\perp})$ for each n in $\mathbb{N}$ and

$$g = \sum_{m=0}^{\infty} T_2^m z_m,$$

where $z_m \in \bigcap_{k=0}^{\infty} T_1^k(R_2^{\perp})$ for each m in $\mathbb{N}$. Therefore,

$$
\begin{aligned}
(f,g)_{\mathcal{H}} &= \left(\sum_{n=0}^{\infty} T_1^n y_n, \sum_{m=0}^{\infty} T_2^m z_m \right)_{\mathcal{H}} \\
&= \sum_{n=0}^{\infty} \sum_{m=0}^{\infty} (T_1^n y_n, T_2^m z_m)_{\mathcal{H}}.
\end{aligned}
$$

Now, for each m in $\mathbb{N}$, $y_n = T_2^m y_{n,m}$, where $y_{n,m} \in R_1^{\perp}$. Similarly, for each n in $\mathbb{N}$, $z_m = T_1^{n+1} z_{m,n+1}$, where $z_{m,n+1} \in R_2^{\perp}$. Therefore,

$$(T_1^n y_n, T_2^m z_m)_{\mathcal{H}} = (T_1^n T_2^m y_{n,m}, T_2^m T_1^{n+1} z_{m,n+1})_{\mathcal{H}} = (y_{n,m}, T_1 z_{m,n+1})_{\mathcal{H}} = 0.$$

Therefore, $(f,g)_{\mathcal{H}} = 0$ and so $B \perp D$, as desired.

We summarize our work in this section in the following theorem.

Theorem 3.4.9 (Halmos Fourfold Decomposition). *Let T_1 and T_2 be doubly commuting isometries in $\mathcal{L}(\mathcal{H})$.*

1. *There is a unique decomposition*

$$\mathcal{H} = \mathcal{H}_{ss} \oplus \mathcal{H}_{su} \oplus \mathcal{H}_{us} \oplus \mathcal{H}_{uu},$$

   where $\mathcal{H}_{ss}$, $\mathcal{H}_{su}$, $\mathcal{H}_{us}$ and $\mathcal{H}_{uu}$ are orthogonal subspaces that reduce T_1 and T_2, with

   - *$T_1|_{\mathcal{H}_{ss}}$ is a shift operator on $\mathcal{H}_{ss}$,*
   - *$T_2|_{\mathcal{H}_{ss}}$ is a shift operator on $\mathcal{H}_{ss}$,*
   - *$T_1|_{\mathcal{H}_{su}}$ is a shift operator on $\mathcal{H}_{su}$,*
   - *$T_2|_{\mathcal{H}_{su}}$ is a unitary operator on $\mathcal{H}_{su}$,*
   - *$T_1|_{\mathcal{H}_{us}}$ is a unitary operator on $\mathcal{H}_{us}$,*

- $T_2|_{\mathcal{H}_{us}}$ is a shift operator on $\mathcal{H}_{us}$,
- $T_1|_{\mathcal{H}_{uu}}$ is a unitary operator on $\mathcal{H}_{uu}$, and
- $T_2|_{\mathcal{H}_{uu}}$ is a unitary operator on $\mathcal{H}_{uu}$.

2. Define $R_1^\perp = \mathcal{H} \ominus T_1\mathcal{H}$ and $R_2^\perp = \mathcal{H} \ominus T_2\mathcal{H}$. Then,

$$\mathcal{H}_{ss} = \sum_{n=0}^{\infty}\sum_{m=0}^{\infty} \oplus T_1^n T_2^m \left(R_1^\perp \cap R_2^\perp \right)$$

$$\mathcal{H}_{su} = \sum_{n=0}^{\infty} \oplus T_1^n \left(\bigcap_{m=0}^{\infty} T_2^m(R_1^\perp) \right)$$

$$\mathcal{H}_{us} = \sum_{n=0}^{\infty} \oplus T_2^n \left(\bigcap_{m=0}^{\infty} T_1^m(R_2^\perp) \right)$$

$$\mathcal{H}_{uu} = \bigcap_{m,n=0}^{\infty} T_1^m T_2^n(\mathcal{H}).$$

3.5   The Doubly Commuting Condition

In the last section, we introduced the doubly commuting condition for operators. Recall that two operators T_1 and T_2 in $\mathcal{L}(\mathcal{H})$ are called **doubly commuting** if T_1 commutes with T_2 and T_1 commutes with T_2^*. It is straightforward to see that T_1 commuting with T_2^* is equivalent to T_2 commuting with T_1^*. Under this doubly commuting condition, we proved the Halmos Fourfold Decomposition. In this section, we look at some conditions that are equivalent to the doubly commuting condition. One of the conditions occurs in the study of weakly stationary random fields.

Before we begin our study of these equivalent conditions, we start by making an observation that will be used later. In our studies, the operators of interest will always be isometries. Therefore, let us suppose that T_1 and T_2 are isometries. Suppose further that one of these isometries is unitary. Without loss of generality, let us suppose it is T_2. We now show that if T_1 and T_2 commute, then T_1 and T_2 doubly commute. To see this, note that if T_1 and T_2 commute, then so must T_1^* and T_2^*. This can be seen by applying the adjoint to each side of the equation $T_1 T_2 = T_2 T_1$, which we know is true since T_1 and T_2 commute. Now, observe that $T_1^* = T_1^* I = T_1^* T_2^* T_2 = T_2^* T_1^* T_2$. Therefore, $T_2 T_1^* = T_2 T_2^* T_1^* T_2 = I T_1^* T_2 = T_1^* T_2$. Here, we used the fact that since T_2 is unitary, $T_2^* T_2 = T_2 T_2^* = I$.

Now, let us consider those aforementioned equivalent conditions. Let $\mathcal{M}$ be a subspace of $\mathcal{H}$. We write $P_{\mathcal{M}}$ to denote the orthogonal projection of $\mathcal{H}$ onto $\mathcal{M}$. We will say T_1 and T_2 are **Radlow commuting** if T_1 commutes with T_2, T_1

commutes with $P_{T_2(\mathcal{H})}$ and T_2 commutes with $P_{T_1(\mathcal{H})}$. James Radlow, in [30], studied the closed ideals of $H^2(\mathbb{T}^2)$ invariant under the operators of multiplication by the coordinate functions and satisfying these commuting conditions. Radlow actually did his study in the more general setting of $H^2(\mathbb{T}^n)$.

Note that if T_j, $j = 1, 2$ are isometries, then $P_{T_j(\mathcal{H})} = T_j T_j^*$, $j = 1, 2$. From this observation, it is straightforward to see that the doubly commuting condition implies the Radlow commuting condition. We will show that the Radlow commuting condition and the doubly commuting condition are equivalent. To see this, suppose that T_1 and T_2 satisfy the Radlow commuting condition. Then, by definition, $T_1 P_{T_2(\mathcal{H})} = P_{T_2(\mathcal{H})} T_1$, which because of our observation above may be written as $T_1 T_2 T_2^* = T_2 T_2^* T_1$. Because T_1 and T_2 commute, we may write this as $T_2 T_1 T_2^* = T_2 T_2^* T_1$. Now multiplying both sides of this equation on the left by T_2^* gives us $T_1 T_2^* = T_2^* T_1$. So, we see that T_1 and T_2 doubly commute. The equivalence of these two conditions verifies the equivalence of the papers of Mandrekar [25] and Radlow [30].

We now look at two other conditions, they both appear in the literature on weakly stationary random fields. In the literature, these conditions are called the strong commuting property and the weak commuting property. Here, as before, we will assume that T_1 and T_2 are isometries. Let $p(n, m) = P_{T_1^n T_2^m(\mathcal{H})} = T_1^n T_2^m T_2^{*m} T_1^{*n}$, $p_1(n) = P_{T_1^n(\mathcal{H})} = T_1^n T_1^{*n}$, and $p_2(m) = P_{T_2^m(\mathcal{H})} = T_2^m T_2^{*m}$. We say that T_1 and T_2 have the **strong commuting property** if T_1 and T_2 commute and $p(n, m) = p_1(n) p_2(m)$ and we say that T_1 and T_2 have the **weak commuting property** if T_1 and T_2 commute and $p_1(n) p_2(m) = p_2(m) p_1(n)$. It is straightforward to see that the strong commuting property implies the weak commuting property. This is simply because $p_1(n) p_2(m) = p(n, m) = p(n, m)^* = (p_1(n) p_2(m))^* = p_2(m)^* p_1(n)^* = p_2(m) p_1(n)$. On the other hand, the weak commuting property does not imply the strong commuting property. To see this, consider the following example, which was inspired by Example 1 in [14]. Let $S = \{(m, n) \in \mathbb{Z}^2 : m \geq 0, n \geq 0, m + n > 0\}$ and let H_S be a Hilbert space with orthonormal basis $\{e_{m,n} : (m, n) \in S\}$. Now, define isometries T_1 and T_2 on H_S by $T_1 e_{i,j} = e_{i+1,j}$ and $T_2 e_{i,j} = e_{i,j+1}$. Let $f \in H_S$. Then,

$$f = \sum_{k=1}^{\infty} \sum_{l=1}^{\infty} \hat{f}(k, l) e_{k,l} + \sum_{k=1}^{\infty} \hat{f}(k, 0) e_{k,0} + \sum_{l=1}^{\infty} \hat{f}(0, l) e_{0,l}.$$

It follows that

$$p_1(n) p_2(m) f = \sum_{k=n}^{\infty} \sum_{l=m}^{\infty} \hat{f}(k, l) e_{k,l} = p_2(m) p_1(n) f.$$

That is, T_1 and T_2 have the weak commuting property. However,

$$p(n, m) f = \sum_{k=n+1}^{\infty} \sum_{l=m+1}^{\infty} \hat{f}(k, l) e_{k,l} + \sum_{k=n+1}^{\infty} \hat{f}(k, m) e_{k,m}$$

$$+ \sum_{l=m+1}^{\infty} \hat{f}(n, l) e_{n,l} \neq p_1(n) p_2(m) f.$$

Therefore, T_1 and T_2 do not have the strong commuting property.

We now show that the strong commuting property is equivalent to the doubly commuting condition. Suppose that T_1 and T_2 have the strong commuting property. Then, $T_1^n T_2^m T_2^{*m} T_1^{*n} = T_1^n T_1^{*n} T_2^m T_2^{*m}$. Now, using the commutativity of T_1 and T_2, we may rewrite the last equation as $T_1^n T_2^m T_1^{*n} T_2^{*m} = T_1^n T_1^{*n} T_2^m T_2^{*m}$. Finally, multiplying on the left, of both sides of the equation, with T_1^{*n} and the right with T_2^m, we get $T_2^m T_1^{*n} = T_1^{*n} T_2^m$. Since this is valid for all $m, n \geq 1$, we see that $T_2 T_1^* = T_1^* T_2$ and so T_1 and T_2 doubly commute. The other direction is straightforward and we leave it to the reader.

3.6   Invariant Subspaces of $L^2(\mathbb{T}^2)$

Let $\mathbb{T}^2 = \mathbb{T} \times \mathbb{T}$. We will write $L^2(\mathbb{T}^2)$ to denote the collection of all Lebesgue measurable functions $f : \mathbb{T}^2 \to \mathbb{C}$ such that

$$\int_{[-\pi,\pi)} \int_{[-\pi,\pi)} \left| f(e^{i\lambda}, e^{i\theta}) \right|^2 d\sigma^2(e^{i\lambda}, e^{i\theta}) < \infty.$$

As always, we will identify functions that are equal $[\sigma^2]$-a.e., where σ^2 denotes normalized Lebesgue measure on $\mathbb{T}^2$. Some important subspaces of $L^2(\mathbb{T}^2)$ that will come into play are

$$H^2(\mathbb{T}^2) = \overline{span} \left\{ e^{im\lambda} e^{in\theta} : m, n \geq 0 \right\},$$

$$H_1^2(\mathbb{T}^2) = \overline{span} \left\{ e^{im\lambda} e^{in\theta} : m \geq 0, n \in \mathbb{Z} \right\},$$

$$H_2^2(\mathbb{T}^2) = \overline{span} \left\{ e^{im\lambda} e^{in\theta} : n \geq 0, m \in \mathbb{Z} \right\},$$

$$L_\theta^2(\mathbb{T}) = \overline{span} \{ e^{in\theta} : n \in \mathbb{Z} \},$$

and

$$L_\lambda^2(\mathbb{T}) = \overline{span} \{ e^{in\lambda} : n \in \mathbb{Z} \}.$$

Let $S_1 : L^2(\mathbb{T}^2) \to L^2(\mathbb{T}^2)$ be the linear operator defined by

$$S_1(f)(e^{i\lambda}, e^{i\theta}) = e^{i\lambda} f(e^{i\lambda}, e^{i\theta})$$

and let $S_2 : L^2(\mathbb{T}^2) \to L^2(\mathbb{T}^2)$ be the linear operator defined by

$$S_2(f)(e^{i\lambda}, e^{i\theta}) = e^{i\theta} f(e^{i\lambda}, e^{i\theta}).$$

It is straightforward to verify that S_1 and S_2 are isometries in $\mathcal{L}(L^2(\mathbb{T}^2))$. A subspace $\mathcal{M}$ of $L^2(\mathbb{T}^2)$ is said to be S-invariant if $\mathcal{M}$ is S_1-invariant and S_2-invariant. Henceforth, if $\mathcal{M}$ is S-invariant, $R_1^\perp = \mathcal{M} \ominus S_1(\mathcal{M})$ and $R_2^\perp = \mathcal{M} \ominus S_2(\mathcal{M})$.

We shall call a subspace $\mathcal{M}$ of $L^2(\mathbb{T}^2)$ doubly invariant if $\mathcal{M}$ is S-invariant,

$S_1|\mathcal{M}$ is unitary on $\mathcal{M}$ and $S_2|\mathcal{M}$ is unitary on $\mathcal{M}$. That is, $S_1(\mathcal{M}) = \mathcal{M}$ and $S_2(\mathcal{M}) = \mathcal{M}$. The following theorem is an analog of Wiener's Theorem proved by P. Ghatage and V. Mandrekar (see [6]). In fact, all of these results can be found in [6] and more generally, for the weighted case, see the work of R. Cheng in [4].

Theorem 3.6.1. *Every doubly invariant subspace of $L^2(\mathbb{T}^2)$ is of the form $1_E L^2(\mathbb{T}^2)$, where E is a measurable subset of $\mathbb{T}^2$ and 1_E is the indicator function of E.*

Proof: We will leave it to the reader to verify that every subspace of the form $1_E L^2(\mathbb{T}^2)$, where E is a measurable subset of $\mathbb{T}^2$, is a doubly invariant subspace of $L^2(\mathbb{T}^2)$. We will prove the other direction.

Let $\mathbf{1}$ denote the function in $L^2(\mathbb{T}^2)$ that takes the value 1 at every point of $\mathbb{T}^2$. Let $q = P_\mathcal{M}\mathbf{1}$, the projection of $\mathbf{1}$ onto $\mathcal{M}$. Then, $\mathbf{1} - q$ is orthogonal to $\mathcal{M}$ and hence orthogonal to $S_1^m S_2^n q$ for all (m, n) in $\mathbb{Z}^2$. It follows from this that $q - |q|^2 = 0$ $[\sigma^2]$-a.e. Therefore, q takes values 0 and 1 $[\sigma^2]$-a.e. Let E denote the subset of $\mathbb{T}^2$ where q takes the value 1. This set is Lebesgue measurable and $q = 1_E$ $[\sigma^2]$-a.e. We will now show that $\mathcal{M} = qL^2(\mathbb{T}^2)$. Since q is in $\mathcal{M}$ and $\mathcal{M}$ is doubly invariant, it follows that $qL^2(\mathbb{T}^2) \subseteq \mathcal{M}$. If this inclusion is strict, then one can find a function in $\mathcal{M} \ominus qL^2(\mathbb{T}^2)$ that is not identically zero. Let us call this function g. Therefore, g is orthogonal to $S_1^m S_2^n q$ for all $(m, n) \in \mathbb{Z}^2$ and hence, $\bar{g}q = 0$ $[\sigma^2]$-a.e. We also observe that since $\mathbf{1} - q$ is orthogonal to $\mathcal{M}$ and $g \in \mathcal{M}$, then $\mathbf{1} - q$ is orthogonal to $S_1^m S_2^n g$ for all $(m, n) \in \mathbb{Z}^2$ and hence $(1 - q)\bar{g} = 0$ $[\sigma^2]$-a.e. Therefore, $g = \bar{g}q + (1 - q)\bar{g} = 0$ $[\sigma^2]$-a.e. Therefore, the containment must not be strict and $\mathcal{M} = qL^2(\mathbb{T}^2)$ as desired. $\square$

Note: Although we did not make use of it in our proof, because at least one of the $S_1|\mathcal{M}$ and $S_2|\mathcal{M}$ are unitary, $S_1|\mathcal{M}$ and $S_2|\mathcal{M}$ doubly commute on $\mathcal{M}$. So, $\mathcal{M} = \mathcal{M}_{uu}$, using the notation from our section on the Halmos Fourfold Decomposition, and has the form given in the above theorem.

Now, let us take a look at another scenario. This time, let us consider the case when $S_1|\mathcal{M}$ is not unitary on $\mathcal{M}$ and $S_2|\mathcal{M}$ is unitary on $\mathcal{M}$. That is, $S_1(\mathcal{M}) \subsetneq \mathcal{M}$ and $S_2(\mathcal{M}) = \mathcal{M}$. Under these conditions, we get the following result.

Theorem 3.6.2. *Every S-invariant subspace $\mathcal{M}$ of $L^2(\mathbb{T}^2)$ for which $S_1(\mathcal{M}) \subsetneq \mathcal{M}$ and $S_2(\mathcal{M}) = \mathcal{M}$ is of the form*

$$\left(\sum_{j=1}^{\infty} \oplus u_j(e^{i\lambda}, e^{i\theta}) 1_{\mathbb{T} \times K_j} H_1^2(\mathbb{T}^2) \right) \bigoplus 1_E L^2(\mathbb{T}^2),$$

where the u_j's are unimodular, the K_j's are measurable subsets of $\mathbb{T}$ with the property that $\sigma(K_j \cap K_l) = 0$ for all $j \neq l$, where σ denotes normalized Lebesgue measure on $\mathbb{T}$, and E is a measurable subset of $\mathbb{T}^2$ with the property that $\sigma^2\left(E \cap \left(\mathbb{T} \times \left(\cup_{j=1}^{\infty} K_j\right)\right)\right) = 0$.

Note: Since $S_2|\mathcal{M}$ is unitary on $\mathcal{M}$ it follows that $S_1|\mathcal{M}$ and $S_2|\mathcal{M}$ doubly commute on $\mathcal{M}$. Our proof follows the approach given in [4].

Proof: By the Halmos Decomposition for $S_1|\mathcal{M}$, we get that

$$\mathcal{M} = \left(\sum_{n=0}^{\infty} \oplus S_1^n(R_1^{\perp}) \right) \oplus \left(\bigcap_{n=0}^{\infty} S_1^n(\mathcal{M}) \right).$$

Since $S_1|\mathcal{M}$ and $S_2|\mathcal{M}$ doubly commute on $\mathcal{M}$, we also get that

$$\mathcal{M} = \left(\sum_{n=0}^{\infty} \oplus S_1^n \left(\bigcap_{m=0}^{\infty} S_2^m(R_1^{\perp}) \right) \right) \oplus \left(\bigcap_{n=0}^{\infty} S_1^n(\mathcal{M}) \right),$$

by the Halmos Fourfold Decomposition. Note that since $S_2(\mathcal{M}) = \mathcal{M}$, $R_2^{\perp} = \{0\}$, which causes two of the four parts of the decomposition to reduce to $\{0\}$. It follows that $\bigcap_{m=0}^{\infty} S_2^m(R_1^{\perp}) = R_1^{\perp}$ and so $S_2(R_1^{\perp}) = R_1^{\perp}$.

Let $e_1 \in R_1^{\perp}$ and let $E_1 = \overline{span}\{S_2^m(e_1) : m \in \mathbb{Z}\}$, which is a subspace of $R_1^{\perp}$. Now, let $e_2 \in R_1^{\perp} \ominus E_1$ and let $E_2 = \overline{span}\{S_2^m(e_2) : m \in \mathbb{Z}\}$, which is a subspace of $R_1^{\perp} \ominus E_1$. Continuing in this manner, we get a sequence of elements $(e_n)_n$ and sequence of subspaces $(E_n)_n$ with the properties that $e_n \in R_1^{\perp} \ominus (E_1 \vee E_2 \vee \cdots \vee E_{n-1})$ and $E_n = \overline{span}\{S_2^m(e_n) : m \in \mathbb{Z}\}$, which is a subspace of $R_1^{\perp} \ominus (E_1 \vee E_2 \vee \cdots \vee E_{n-1})$. Observe that by construction,

$$(S_1^n e_j, S_2^m e_k)_{L^2(\mathbb{T}^2)} = \begin{cases} 0 & \text{if } k \neq j, n \geq 0, m \in \mathbb{Z} \\ 0 & \text{if } k = j, n > 0, m \in \mathbb{Z} \end{cases}$$

and

$$(S_2^m e_j, S_1^n e_k)_{L^2(\mathbb{T}^2)} = \begin{cases} 0 & \text{if } k \neq j, n \geq 0, m \in \mathbb{Z} \\ 0 & \text{if } k = j, n > 0, m \in \mathbb{Z} \end{cases}.$$

Therefore, we may conclude that for $k \neq j$, $e_j \bar{e}_k = 0$, $[\sigma^2]$-a.e. and for all j, $|e_j(e^{i\lambda}, e^{i\theta})|^2 = f_j(e^{i\theta})$, for some $f_j \in L^1(\mathbb{T})$. Define $\mathfrak{S}_j = \{(e^{i\lambda}, e^{i\theta}) : e_j(e^{i\lambda}, e^{i\theta}) \neq 0\}$. Then, $(\mathfrak{S}_j)_j$ is a sequence of measurable subsets of $\mathbb{T}^2$ with the property that $\sigma^2(\mathfrak{S}_j \cap \mathfrak{S}_k) = 0$ for all $j \neq k$ and each $\mathfrak{S}_j$ is of the form $\mathbb{T} \times K_j$, where K_j is a measurable subset of $\mathbb{T}$. Furthermore, $R_1^{\perp} = \sum_{j=1}^{\infty} \oplus E_j$.

Now, using the polar decomposition we get that $e_j(e^{i\lambda}, e^{i\theta}) = u_j(e^{i\lambda}, e^{i\theta})g_j(e^{i\theta})$, where u has modulus one, and $g = |e_j| = \sqrt{f_j}$. From this decomposition, we see that $E_j = u_j(e^{i\lambda}, e^{i\theta})\mathbf{1}_{\mathbb{T} \times K_j} L_{\theta}^2(\mathbb{T})$. Hence,

$$R_1^{\perp} = \sum_{j=1}^{\infty} \oplus u_j(e^{i\lambda}, e^{i\theta})\mathbf{1}_{\mathbb{T} \times K_j} L_{\theta}^2(\mathbb{T})$$

and

$$\sum_{n=0}^{\infty} \oplus S_1^n(R_1^{\perp}) = \sum_{n=0}^{\infty} \oplus S_1^n \left(\sum_{j=1}^{\infty} \oplus u_j(e^{i\lambda}, e^{i\theta})\mathbf{1}_{\mathbb{T} \times K_j} L_{\theta}^2(\mathbb{T}) \right)$$

$$= \sum_{j=1}^{\infty} \oplus u_j(e^{i\lambda}, e^{i\theta})\mathbf{1}_{\mathbb{T} \times K_j} H_1^2(\mathbb{T}^2).$$

It remains to show that $\bigcap_{n=0}^{\infty} S_1^n(\mathcal{M}) = \mathbf{1}_E L^2(\mathbb{T}^2)$, where E is a measurable subset of $\mathbb{T}^2$, with the property that $\sigma^2\left(E \cap \left(\mathbb{T} \times \left(\cup_{j=1}^{\infty} K_j\right)\right)\right) = 0$. This follows from the fact that each part of the Halmos Fourfold Decomposition reduces both S_1 and S_2. Therefore, $\bigcap_{n=0}^{\infty} S_1^n(\mathcal{M})$ is an S-invariant subspace of $L^2(\mathbb{T}^2)$ where the restriction of both S_1 and S_2 is unitary. Therefore, by the previous theorem, $\bigcap_{n=0}^{\infty} S_1^n(\mathcal{M}) = \mathbf{1}_E L^2(\mathbb{T}^2)$, where E is a measurable subset of $\mathbb{T}^2$. To see that E has the property that $\sigma^2\left(E \cap \left(\mathbb{T} \times \left(\cup_{j=1}^{\infty} K_j\right)\right)\right) = 0$, we just need to recall that our decomposition is orthogonal and the desired conclusion follows. $\square$

By symmetry, we get the following result.

Theorem 3.6.3. *Every S-invariant subspace $\mathcal{M}$ of $L^2(\mathbb{T}^2)$ for which $S_1(\mathcal{M}) = \mathcal{M}$ and $S_2(\mathcal{M}) \subsetneq \mathcal{M}$ is of the form*

$$\left(\sum_{j=1}^{\infty} \oplus u_j(e^{i\lambda}, e^{i\theta}) \mathbf{1}_{K_j \times \mathbb{T}} H_2^2(\mathbb{T}^2)\right) \bigoplus \mathbf{1}_E L^2(\mathbb{T}^2),$$

where the u_j's are unimodular, the K_j's are measurable subsets of $\mathbb{T}$ with the property that $\sigma(K_j \cap K_l) = 0$ for all $j \neq l$ and E is a measurable subset of $\mathbb{T}^2$ with the property that $\sigma^2\left(E \cap \left(\left(\cup_{j=1}^{\infty} K_j\right) \times \mathbb{T}\right)\right) = 0$.

We are left to consider the case when $S_1(\mathcal{M}) \subsetneq \mathcal{M}$ and $S_2(\mathcal{M}) \subsetneq \mathcal{M}$. Because neither $S_1|\mathcal{M}$ nor $S_2|\mathcal{M}$ is unitary on $\mathcal{M}$, these operators, in general, may not be doubly commuting on $\mathcal{M}$. A complete description of all such S-invariant subspaces is not known. There are many cases where it is known. In particular, a description is known when $S_1|\mathcal{M}$ and $S_2|\mathcal{M}$ doubly commute on $\mathcal{M}$.

Theorem 3.6.4. *Suppose that $\mathcal{M}$ is an S-invariant subspace of $L^2(\mathbb{T}^2)$. The following are equivalent.*

    1. *$S_1|\mathcal{M}$ and $S_2|\mathcal{M}$ doubly commute on $\mathcal{M}$ and $R_1^{\perp} \cap R_2^{\perp} = \{0\}$.*

    2. *$S_1|\mathcal{M}$ is unitary on $\mathcal{M}$ or $S_2|\mathcal{M}$ is unitary on $\mathcal{M}$.*

Proof: 2 implies 1 is straightforward since, as we mentioned above, if either $S_1|\mathcal{M}$ is unitary on $\mathcal{M}$ or $S_2|\mathcal{M}$ is unitary on $\mathcal{M}$, then $S_1|\mathcal{M}$ and $S_2|\mathcal{M}$ doubly commute on $\mathcal{M}$. Also, if say $S_1|\mathcal{M}$ is unitary on $\mathcal{M}$, then $R_1^{\perp} = \{0\}$ and so $R_1^{\perp} \cap R_2^{\perp} = \{0\}$. Similarly, if $S_2|\mathcal{M}$ is unitary on $\mathcal{M}$.

    1 implies 2 requires a bit more work. Since $S_1|\mathcal{M}$ and $S_2|\mathcal{M}$ doubly commute on $\mathcal{M}$, $\mathcal{M}$ has the Halmos Fourfold Decompostion. Since $R_1^{\perp} \cap R_2^{\perp} = \{0\}$ at most three of the parts of the decomposition are nonzero. Therefore, $\mathcal{M}$

has the following form.

$$\mathcal{M} = \left(\sum_{n=0}^{\infty} \oplus S_1^n \left(\bigcap_{m=0}^{\infty} S_2^m(R_1^{\perp})\right)\right) \oplus \left(\sum_{n=0}^{\infty} \oplus S_2^n \left(\bigcap_{m=0}^{\infty} S_1^m(R_2^{\perp})\right)\right)$$
$$\oplus \left(\bigcap_{m,n=0}^{\infty} S_1^m S_2^n(\mathcal{M})\right).$$

We will show that at most two parts of the decomposition are nonzero, which will give us our desired result. To avoid so much writing, I will use $\mathcal{M}_{su}$ to denote the first part of the above decomposition, $\mathcal{M}_{us}$ to denote the second part of the above decomposition, and $\mathcal{M}_{uu}$ to denote the third and final part of the above decomposition. By our work above, we know that

$$\mathcal{M}_{su} = \sum_{j=1}^{\infty} \oplus u_j(e^{i\lambda}, e^{i\theta}) \mathbf{1}_{\mathbb{T} \times K_j} H_1^2(\mathbb{T}^2),$$

$$\mathcal{M}_{us} = \sum_{j=1}^{\infty} \oplus v_j(e^{i\lambda}, e^{i\theta}) \mathbf{1}_{N_j \times \mathbb{T}} H_2^2(\mathbb{T}^2),$$

and

$$\mathcal{M}_{uu} = \mathbf{1}_E L^2(\mathbb{T}^2),$$

where the u_j's are unimodular, the v_j's are unimodular, the K_j's are measurable subsets of $\mathbb{T}$ with the property that $\sigma(K_j \cap K_l) = 0$ for all $j \neq l$, the N_j's are measurable subsets of $\mathbb{T}$ with the property that $\sigma(N_j \cap N_l) = 0$ for all $j \neq l$, and because our decomposition is orthogonal, there are extra restrictions imposed on these sets. The orthogonality forces $\sigma(K_j) \cdot \sigma(N_l) = 0$ for all j, l. Therefore, if there exists a single K_j such that $\sigma(K_j) \neq 0$, then $\sigma(N_j) = 0$ for all j. Hence $\mathcal{M}_{us} = \{0\}$. It then follows from the Halmos Fourfold Decomposition that $S_2|\mathcal{M}$ is unitary on $\mathcal{M}$. If on the other hand, $\sigma(K_j) = 0$ for all j, then $\mathcal{M}_{su} = \{0\}$ and $S_1|\mathcal{M}$ is unitary on $\mathcal{M}$. There are also conditions on E, but they are not needed for this proof. $\square$

Corollary 3.6.1. *Suppose that $\mathcal{M}$ is an S-invariant subspace of $L^2(\mathbb{T}^2)$. Furthermore, suppose that $S_1|\mathcal{M}$ and $S_2|\mathcal{M}$ doubly commute on $\mathcal{M}$ with $R_1^{\perp} \cap R_2^{\perp} = \{0\}$. Then, $\mathcal{M}$ has the form given in either Theorem 3.6.1, Theorem 3.6.2 or Theorem 3.6.3.*

Theorem 3.6.5. *Suppose that $\mathcal{M}$ is an S-invariant subspace of $L^2(\mathbb{T}^2)$. Furthermore, suppose that $S_1|\mathcal{M}$ and $S_2|\mathcal{M}$ doubly commute on $\mathcal{M}$ with $S_1(\mathcal{M}) \subsetneq \mathcal{M}$ and $S_2(\mathcal{M}) \subsetneq \mathcal{M}$. Then, $\mathcal{M} = \varphi H^2(\mathbb{T}^2)$, where φ is unimodular.*

Proof: By our supposition and Theorem 3.6.4, it follows that $R_1^{\perp} \cap R_2^{\perp} \neq \{0\}$ and that $\mathcal{M}$ has the Halmos Fourfold Decomposition. In the previous theorems, we have analyzed three of the parts in detail. We will begin by looking at the remaining part. We will denote this part by $\mathcal{M}_{ss}$. We know from the Halmos Fourfold Decomposition, it has the form

$$\mathcal{M}_{ss} = \sum_{n=0}^{\infty} \sum_{m=0}^{\infty} \oplus S_1^n S_2^m (R_1^{\perp} \cap R_2^{\perp}).$$

Since $R_1^{\perp} \cap R_2^{\perp} \neq \{0\}$, let $\varphi \in R_1^{\perp} \cap R_2^{\perp}$ with the property that $\|\varphi\|_{L^2(\mathbb{T}^2)} = 1$. By the decomposition, we see that

$$(S_1^n \varphi, S_2^m \varphi)_{L^2(\mathbb{T}^2)} = \begin{cases} 0 & \text{if } n \geq 0, m > 0 \\ 0 & \text{if } n > 0, m \geq 0 \end{cases},$$

$$(S_2^n \varphi, S_1^m \varphi)_{L^2(\mathbb{T}^2)} = \begin{cases} 0 & \text{if } n \geq 0, m > 0 \\ 0 & \text{if } n > 0, m \geq 0 \end{cases},$$

$$(S_1^m S_2^n \varphi, \varphi)_{L^2(\mathbb{T}^2)} = \begin{cases} 0 & \text{if } n \geq 0, m > 0 \\ 0 & \text{if } n > 0, m \geq 0 \end{cases},$$

$$(\varphi, S_1^m S_2^n \varphi)_{L^2(\mathbb{T}^2)} = \begin{cases} 0 & \text{if } n \geq 0, m > 0 \\ 0 & \text{if } n > 0, m \geq 0 \end{cases},$$

and

$$(\varphi, \varphi)_{L^2(\mathbb{T}^2)} = 1.$$

Therefore, $|\varphi|^2 = 1$, $[\sigma^2]$-a.e. Therefore, φ is unimodular. Now, let $\psi \in R_1^{\perp} \cap R_2^{\perp}$ with $\psi \perp \varphi$. Then,

$$(S_1^n \varphi, S_2^m \psi)_{L^2(\mathbb{T}^2)} = \begin{cases} 0 & \text{if } n \geq 0, m > 0 \\ 0 & \text{if } n > 0, m \geq 0 \end{cases},$$

$$(S_2^n \varphi, S_1^m \psi)_{L^2(\mathbb{T}^2)} = \begin{cases} 0 & \text{if } n \geq 0, m > 0 \\ 0 & \text{if } n > 0, m \geq 0 \end{cases},$$

$$(S_1^m S_2^n \varphi, \psi)_{L^2(\mathbb{T}^2)} = \begin{cases} 0 & \text{if } n \geq 0, m > 0 \\ 0 & \text{if } n > 0, m \geq 0 \end{cases},$$

$$(\varphi, S_1^m S_2^n \psi)_{L^2(\mathbb{T}^2)} = \begin{cases} 0 & \text{if } n \geq 0, m > 0 \\ 0 & \text{if } n > 0, m \geq 0 \end{cases},$$

and

$$(\varphi, \psi)_{L^2(\mathbb{T}^2)} = 0.$$

Therefore, $\varphi \overline{\psi} = 0$, $[\sigma^2]$-a.e. Since φ is unimodular, it follows that $\psi = 0$, $[\sigma^2]$-a.e. It follows that $R_1^{\perp} \cap R_2^{\perp}$ is spanned by the unimodular function φ. Therefore, $\mathcal{M}_{ss} = \varphi H^2(\mathbb{T}^2)$. We will be done when we show that the remaining parts of the Halmos Fourfold Decomposition are $\{0\}$. First note that φ does not vanish on a set of positive measure. Then, using the fact that our decomposition is orthogonal, we get that $\sigma^2(N_j \times \mathbb{T}) = 0$ for all j, $\sigma^2(\mathbb{T} \times K_j) = 0$ for all j, and $\sigma^2(E) = 0$. Therefore, $\mathcal{M}_{su} = \mathcal{M}_{us} = \mathcal{M}_{uu} = \{0\}$ and our theorem follows. $\square$

3.7 Invariant Subspaces of $H^2(\mathbb{T}^2)$

We now consider nonzero S-invariant subspaces $\mathcal{M}$ of $H^2(\mathbb{T}^2)$. Since $H^2(\mathbb{T}^2)$ is a subspace of $L^2(\mathbb{T}^2)$, it follows that $\mathcal{M}$ is also a nonzero S-invariant subspace of $L^2(\mathbb{T}^2)$. By our work in the previous section, we know that if $S_1|\mathcal{M}$ and $S_2|\mathcal{M}$ doubly commute on $\mathcal{M}$, then $\mathcal{M}$ has one of the forms given in Theorems 3.6.1, 3.6.2, 3.6.3, or 3.6.5. We will show that nonzero S-invariant subspaces of $H^2(\mathbb{T}^2)$ cannot be of the forms given in Theorems 3.6.1, 3.6.2, or 3.6.3.

Suppose first that $\mathcal{M}$ has the form given in Theorem 3.6.1. That is,

$$\mathcal{M} = 1_E L^2(\mathbb{T}^2), \tag{3.3}$$

where E is a measurable subset of $\mathbb{T}^2$. Since $\mathcal{M} \subset H^2(\mathbb{T}^2)$, it follows that no member of $\mathcal{M}$ can vanish on a set of positive measure. Therefore, $\sigma^2(E) = 1$ and so $\mathcal{M} = L^2(\mathbb{T}^2)$. This is a contradiction since $H^2(\mathbb{T}^2) \subsetneq L^2(\mathbb{T}^2)$. So, that eliminates that possibility.

Now, suppose that $\mathcal{M}$ has the form given in Theorem 3.6.2. That is,

$$\mathcal{M} = \left(\sum_{j=1}^{\infty} \oplus u_j(e^{i\lambda}, e^{i\theta}) 1_{\mathbb{T} \times K_j} H_1^2(\mathbb{T}^2) \right) \bigoplus 1_E L^2(\mathbb{T}^2), \tag{3.4}$$

where the u_j's are unimodular, the K_j's are measurable subsets of $\mathbb{T}$ with the property that $\sigma(K_j \cap K_l) = 0$ for all $j \neq l$, and E is a measurable subset of $\mathbb{T}^2$ with the property that $\sigma^2\left(E \cap \left(\mathbb{T} \times \left(\cup_{j=1}^{\infty} K_j\right)\right)\right) = 0$. Again, because $\mathcal{M} \subset H^2(\mathbb{T}^2)$, no member of $\mathcal{M}$ can vanish on a set of positive measure. It then follows either $\sigma^2(E) = 0$ and $\sigma^2\left(\left(\mathbb{T} \times \left(\cup_{j=1}^{\infty} K_j\right)\right)\right) = 1$, or $\sigma^2\left(\left(\mathbb{T} \times \left(\cup_{j=1}^{\infty} K_j\right)\right)\right) = 0$ and $\sigma^2(E) = 1$. We know from above that the only true possibility is when $\sigma^2(E) = 0$ and $\sigma^2\left(\left(\mathbb{T} \times \left(\cup_{j=1}^{\infty} K_j\right)\right)\right) = 1$. In this case, $\mathcal{M}$ may be written in the form

$$\mathcal{M} = \sum_{j=1}^{\infty} \oplus u_j(e^{i\lambda}, e^{i\theta}) 1_{\mathbb{T} \times K_j} H_1^2(\mathbb{T}^2), \tag{3.5}$$

where the u_j's are unimodular, the K_j's are measurable subsets of $\mathbb{T}$ with the property that $\sigma(K_j \cap K_l) = 0$ for all $j \neq l$, and $\sigma^2\left(\left(\mathbb{T} \times \left(\cup_{j=1}^{\infty} K_j\right)\right)\right) = 1$. It follows from these conditions and the fact that $1 \in H_1^2(\mathbb{T}^2)$, that $u(e^{i\lambda}, e^{i\theta}) = \sum_{j=1}^{\infty} u_j(e^{i\lambda}, e^{i\theta}) 1_{\mathbb{T} \times K_j}$ is in $H^2(\mathbb{T}^2)$ and that $u H_1^2(\mathbb{T}^2) \subset \mathcal{M}$. Therefore, $e^{-in\theta} u \in \mathcal{M}$ for all n in $\mathbb{Z}$, which gives a contradiction, since there exists an n in $\mathbb{Z}$ such that $e^{-in\theta} u \notin H^2(\mathbb{T}^2)$. So, this possibility is also out. By symmetry, the form of $\mathcal{M}$ given by Theorem 3.6.3 is also out. This leaves just one possibility.

Finally, suppose that $\mathcal{M}$ has the form given in Theorem 3.6.5. That is,

$$\mathcal{M} = \varphi H^2(\mathbb{T}^2), \tag{3.6}$$

where φ is unimodular. We only note that since $1 \in H^2(\mathbb{T}^2)$, if follows that φ is also in $H^2(\mathbb{T}^2)$. We recall that functions that are both unimodular and in $H^2(\mathbb{T}^2)$ are called inner functions. So, φ is an inner function. It is straightforward to verify that such a subspace is in $H^2(\mathbb{T}^2)$ and is indeed S-invariant. We summarize our findings in the following theorem.

Theorem 3.7.1. *Suppose that* $\mathcal{M}$ *is a nonzero* S-*invariant subspace of* $H^2(\mathbb{T}^2)$ *with the property that* $S_1|\mathcal{M}$ *and* $S_2|\mathcal{M}$ *doubly commute on* $\mathcal{M}$. *Then,* $\mathcal{M} = \varphi H^2(\mathbb{T}^2)$, *where* φ *is an inner function.*

3.8 Remarks and Related Literature

The Halmos Fourfold Decomposition Theorem was proved by Slociński [39] and separately by Kallianpur and Mandrekar [18]. The work on invariant subspaces of $H^2(\mathbb{T}^2)$ was proved by Agrawal, Clark and Douglas [1] and Mandrekar [25]. Using their ideas, subspaces of $L^2(\mathbb{T}^2)$ were studied by P. Ghatage and Mandrekar [6] using the Halmos Fourfold Decomposition Theorem. Other work on detailed analysis of such subspaces from the analytic point of view was done by Izuchi, Nakazi and Seto [INS].

[INS] Keiji Izuchi, Takahiko Nakazi, Michio Seto, *Backward shift invariant subspaces in the bidisc. III.* Acta Sci. Math. (Szeged) 70 (2004), no. 3-4, 727–749.

4

Applications and Generalizations

4.1 Texture Identification

We represent a texture observed at a point (m, n) by

$$z(m, n) = \sum_{k=1}^{p} \{ C_k \cos(m\lambda_k + n\mu_k) + D_k \sin(m\lambda_k + n\mu_k) \},$$

where $\{(\lambda_k, \mu_k) : k = 1, \cdots, p\}$ are connected with spectrum intensities and $\{(C_k, D_k) : k = 1, \cdots, p\}$ are amplitudes. As the texture is observed under light conditions, what we observe is

$$y(m, n) = z(m, n) + x(m, n), \tag{4.1}$$

where $x(m, n)$ is a random field. We make the assumption that

$$x(m, n) = \sum_{j=-\infty}^{\infty} \sum_{k=-\infty}^{\infty} c(j, k) \, \varepsilon(m - j, n - k),$$

where $c(j, k)$ are constants with $\sum_{j=-\infty}^{\infty} \sum_{k=-\infty}^{\infty} |c(j, k)| < \infty$ and $\{\varepsilon(m, n) : m,$ $n \in \mathbb{Z}\}$ is a double array of independent random fields with $E\varepsilon(m, n) = 0$, $E|\varepsilon(m, n)|^2 = 1$ and $\sup_{m,n} E|\varepsilon(m, n)|^r < \infty$ for some $r > 2$. These conditions imply that $x(m, n)$ is stationary.

The technique we use is motivated by the univariate case studied by Priestley [27], [28] and Whittle [41], [42] using the idea of a periodogram (for the review, see Priestley [29]). The problem of estimating (λ_k, μ_k) and (C_k, D_k) with known p uses maximum-likelihood methods (Kundu and Mitra [20]; Bansal, Hamedani and Zhang [2]; and Rao, Zhao and Zhou [31]). However, in order to identify the texture, one needs to estimate p, which was done by Zhang and Mandrekar [46]. Our presentation is based on this work.

Let us define the periodogram of $\{y(m, n) : m, n \in \mathbb{Z}\}$ as

$$I_N(\lambda, \mu; y) = \frac{1}{(2\pi N)^2} \left| \sum_{m=1}^{N} \sum_{n=1}^{N} y(m, n) e^{-i(m\lambda + n\mu)} \right|^2.$$

Since each term in $z(m, n)$ can be written as

$$C_k \cos(-m\lambda_k - n\mu_k) - D_k \sin(-m\lambda_k - n\mu_k),$$

we assume that $\mu_k \geq 0$ for $k = 1, \cdots, p$ to make the model identifiable and this does not constrain the model. Since $I_N(\lambda, \mu; y) = I_N(-\lambda, -\mu; y)$, we restrict $I_N(\lambda, \mu; \cdot)$ to the set $\Pi = (-\pi, \pi] \times [0, \pi]$.

The basic idea of our proof is to show that the periodograms of x and that of z behave differently as $N \to \infty$. In view of this, we need to use a law of iterated logarithm (LIL) to obtain the asymptotic behaviors of $I_N(\lambda, \mu; x)$. We note that from our general result x is stationary and has spectral density. We denote it by $f(\lambda, \mu)$.

Let us denote by $I_N(\lambda, \mu) = I_N(\lambda, \mu; x)$. Then, we shall establish an upper bound for this periodogram. It is based on the following lemma.

Lemma 4.1.1. *Let $\{\varepsilon(m, n) : m, n \in \mathbb{Z}\}$ be a double array of independent random fields such that for $m, n \in \mathbb{Z}$, $E\varepsilon(m, n) = 0$, $E\varepsilon^2(m, n) = 1$ and $\sup\limits_{m,n} E|\varepsilon(m, n)|^r < \infty$ for some $r > 2$. For $N \geq 1$, let $\{a_N(m, n) : m, n \in \mathbb{Z}\}$ be a double array of constants such that*

$$A_N = \sum_{m=-\infty}^{\infty} \sum_{n=-\infty}^{\infty} a_N^2(m, n) < \infty \quad \text{and} \quad \lim_{N \to \infty} A_N = \infty$$

and

$$\sup_{m,n} a_N^2(m, n) = o\left(A_N (\log A_N)^{-\rho}\right) \quad \text{for all } \rho > 0.$$

Suppose there exist constants $\alpha_j > 0$ and $d > 2/r$ such that for some $M_0 > 0$ and all $N > M \geq M_0$,

$$\sum_{m=-\infty}^{\infty} \sum_{n=-\infty}^{\infty} (a_N(m, n) - a_M(m, n))^2 \leq \left(\sum_{j=M+1}^{N} \alpha_j\right)^d$$

and as $N \to \infty$,

$$\left(\sum_{j=M_0}^{N} \alpha_j\right)^d = O(A_N).$$

Define

$$S_N = \sum_{m=-\infty}^{\infty} \sum_{n=-\infty}^{\infty} a_N(m, n)\, \varepsilon(m, n)$$

and $\sigma_N^2 = \mathrm{Var}(S_N) = A_N$. Then,

$$\limsup_{N \to \infty} \frac{|S_N|}{(2\sigma_N^2 \ln \ln \sigma_N^2)^{1/2}} \leq 1.$$

Remark 1. *This result is a two-dimensional generalization of Theorem 1(i) of T. L. Lai and C. Z. Wei [21] on LIL. The proof is identical and will not be reproduced.*

We shall use this lemma to establish the following theorem.

Theorem 4.1.1. *Let $I_N(\lambda, \mu)$ be the periodogram of $x(m, n)$ defined above, with spectral density $f(\lambda, \mu)$. Then, for each (λ, μ),*

$$\limsup_{N \to \infty} \frac{I_N(\lambda, \mu)}{\ln \ln N} \leq 2f(\lambda, \mu).$$

Furthermore, if in the definition of $x(m, n)$, $E|\varepsilon(m, n)|^r < \infty$ for $r > 3$, then

$$\limsup_{N \to \infty} \frac{\sup_{(\lambda, \mu) \in \Pi} I_N(\lambda, \mu)}{\ln(N^2)} \leq 7\|f\|,$$

where $\|f\| = \sup_{(\lambda, \mu) \in \Pi} f(\lambda, \mu)$.

Proof: Define for $\theta = (\lambda, \mu)$,

$$S_N^+(\theta) = \sum_{m=1}^N \sum_{n=1}^N x(m, n) \left\{\cos(m\lambda + n\mu) + \sin(m\lambda + n\mu)\right\}$$

and

$$S_N^-(\theta) = \sum_{m=1}^N \sum_{n=1}^N x(m, n) \left\{\cos(m\lambda + n\mu) - \sin(m\lambda + n\mu)\right\}.$$

Then

$$\text{Var}(S_N^\pm) = (2\pi N)^2 E(I_N(\theta))$$

$$\pm 2 \, \text{Cov}\left(\sum_{m=1}^N \sum_{n=1}^N x(m, n) \cos(m\lambda + n\mu), \sum_{m=1}^N \sum_{n=1}^N x(m, n) \sin(m\lambda + n\mu)\right).$$

Analogous to the time series case, the periodogram is asymptotically unbiased, and the real part and the imaginary part of the finite Fourier transformation of $x(m, n)$ are asymptotically uncorrelated. Therefore,

$$(2\pi N)^{-2} \, \text{Var}(S_N^\pm) \to f(\lambda, \mu).$$

Since

$$I_N(\theta) = \frac{1}{2(2\pi N)^2} \left((S_N^+)^2 + (S_N^-)^2\right),$$

it suffices to show

$$\limsup_{N \to \infty} \frac{\left|S_N^\pm(\theta)\right|}{\sqrt{2 \, \text{Var}(S_N^\pm(\theta)) \ln \ln \, \text{Var}(S_N^\pm(\theta))}} \leq 1 \text{ for any } \theta \qquad (4.2)$$

and

$$\limsup_{N\to\infty} \frac{\max_\theta \left|S_N^\pm(\theta)\right|}{\sqrt{N^2\ln(N^2)}} \le 2\pi\sqrt{7}\|f\|. \tag{4.3}$$

We only prove (4.2) and (4.3) for S_N^+. Proofs of (4.2) and (4.3) for S_N^- are similar. Let us rewrite $S_N^+(\theta)$ as

$$S_N^+(\theta) = \sum_{j=-\infty}^{\infty}\sum_{k=-\infty}^{\infty} a_N(j,k;\theta)\,\varepsilon(j,k),$$

where

$$a_N(j,k;\theta) = \sum_{m=1}^{N}\sum_{n=1}^{N} c(m-j,n-k)\left\{\cos(m\lambda+n\mu)+\sin(m\lambda+n\mu)\right\}.$$

Then,

$$\|a\| = \sup_{N,j,k,\theta} |a_N(j,k;\theta)| \le 2\sum_{j=-\infty}^{\infty}\sum_{k=-\infty}^{\infty}|c(j,k)| < \infty, \tag{4.4}$$

$$A_N \stackrel{\text{def}}{=} \sum_{j=-\infty}^{\infty}\sum_{k=-\infty}^{\infty} a_N^2(j,k;\theta) = \mathrm{Var}(S_N^+(\theta)) < \infty \text{ and } \frac{A_N}{(2\pi N)^2} \to f(\theta).$$

Note also, for any $N > M > 0$,

$$\sum_{m=-\infty}^{\infty}\sum_{n=-\infty}^{\infty} (a_N(m,n;\theta)-a_M(m,n;\theta))^2 = E(S_N^+(\theta)-S_M^+(\theta))^2$$

$$= \int_{-\pi}^{\pi}\int_{-\pi}^{\pi}\left|\sum_{(m,n)\in D_N}\left\{\cos(m\lambda+n\mu)+\sin(m\lambda+n\mu)\right\}e^{i(m\tilde\lambda+n\tilde\mu)}\right|^2 f(\tilde\lambda,\tilde\mu)\,d\tilde\lambda\,d\tilde\mu$$

$$\le \|f\|\int_{-\pi}^{\pi}\int_{-\pi}^{\pi}\left|\sum_{(m,n)\in D_N}\left\{\cos(m\lambda+n\mu)+\sin(m\lambda+n\mu)\right\}e^{i(m\tilde\lambda+n\tilde\mu)}\right|^2 d\tilde\lambda\,d\tilde\mu$$

$$= \|f\|(2\pi)^2 \sum_{(m,n)\in D_N}\left[\cos(m\lambda+n\mu)+\sin(m\lambda+n\mu)\right]^2$$

$$\le 2(2\pi)^2\|f\|(N^2-M^2) \le \sum_{j=M+1}^{N}\alpha_j,$$

where $D_N = \left\{(m,n)\in\mathbb{Z}^2 : M < m \le N, \text{ or } M < n \le N\right\}, \alpha_j = 4(2\pi)^2\|f\|j$, $j = M+1,\cdots,N$. Applying our lemma with $d = 1 > 2/r$ gives (4.2).

To prove (4.3), we need to truncate $\varepsilon(m,n)$. Let

$$\tilde{\varepsilon}(j,k) = \tilde{\varepsilon}(j,k;\theta) = \varepsilon(j,k)\mathbf{1}_{\{|a_N(j,k;\theta)\,\varepsilon(j,k)| < N/\ln(N^2)\}},$$

$$\tilde{S}_N(\theta) = \sum_{j=-\infty}^{\infty}\sum_{k=-\infty}^{\infty} a_N(j,k;\theta)\,\tilde{\varepsilon}(j,k) \text{ and } S_N^* = \tilde{S}_N(\theta) - E(\tilde{S}_N(\theta)).$$

We will first show, for any $\alpha > 7/2$, that

$$\limsup_{N\to\infty} \frac{\max_\theta |S_N^*(\theta)|}{\sqrt{N^2\ln(N^2)}} \le 2\pi\sqrt{2\alpha}\|f\|. \tag{4.5}$$

For this end, let us show for any fixed θ, and for any $\beta > \alpha > 7/2$, that

$$P\left(|S_N^*(\theta)| > 2\pi\sqrt{2\beta\|f\|N^2\ln(N^2)}\right) \le 2N^{-2\alpha}, \tag{4.6}$$

for all $N > N_0$, where N_0 does not depend on θ.

Since $E\tilde{\varepsilon}^2(j,k) \le E\varepsilon^2(j,k) = 1$,

$$E\left(S_N^{*2}(\theta)\right) \le E\left(S_N^2(\theta)\right) = \int_{-\pi}^{\pi}\int_{-\pi}^{\pi}\left|\sum_{m=1}^{N}\sum_{n=1}^{N} e^{-i(m\lambda+n\mu)}\right|^2$$
$$\times f(\lambda,\mu)\,d\lambda\,d\mu \le (2\pi)^2\|f\|N^2$$

and $\sup_{j,k}|a_N(j,k;\theta)|\,|\tilde{\varepsilon}(j,k) - E\tilde{\varepsilon}(j,k)| \le 2N/\ln(N^2)$, we can apply Lemma 3 (i) of Lai and Wei [21] to $S_N^*(\theta)$ with

$$A = (2\pi)^2\|f\|N^2, c = \frac{1}{\pi\sqrt{\|f\|}\ln(N^2)}, \text{ and } \xi = \sqrt{2\beta\ln(N^2)}$$

to obtain that, for a sufficiently large N such that $c\xi < 1$,

$$P\left(|S_N^*(\theta)| > 2\pi\sqrt{2\beta\|f\|N^2\ln(N^2)}\right) \le$$

$$2\exp\left\{-\beta\ln(N^2)\left(1 - \frac{1}{2\pi\sqrt{\|f\|}\ln(N^2)}\sqrt{2\beta\ln(N^2)}\right)\right\}$$

$$\le 2e^{-\alpha\ln(N^2)} = 2N^{-2\alpha}.$$

Now, choose a q such that $3 < q < (\alpha - 0.5)$ and divide $[-\pi,\pi]\times[-\pi,\pi]$ into $K_N = [N^p]^2$ equal-sized squares, each with a width $2\pi/[N^q]$, where $[N^q]$ is the integer part of N^q. Denote the squares by $\Delta_1, \Delta_2, \cdots, \Delta_{K_n}$ and their centers by $\theta_1, \theta_2, \cdots, \theta_{K_N}$.

Note that $E|\tilde{\varepsilon}(j,k)| \le E|\varepsilon(j,k)| \le 1$ and

$$E\left(\sup_{\theta\in\Delta_k,k}|S_N^*(\theta) - S_N^*(\theta_k)|\right) \le 2\sum_j\sum_k \sup_{\theta\in\Delta_l,l}|a_N(j,k;\theta) - a_N(j,k;\theta_l)|.$$

Since

$$|\cos(m\lambda + n\mu) + \sin(m\lambda + n\mu) - \cos(m\lambda_k + n\mu_k) - \sin(m\lambda_k + n\mu_k)|$$

$$\leq 2|m(\lambda - \lambda_k) + n(\mu - \mu_k)|$$

$$\leq 4\pi\sqrt{2}[N^q]^{-1}\sqrt{m^2 + n^2}, \qquad \forall k \text{ and } (\lambda, \mu) \in \Delta_k,$$

where $\theta_k = (\lambda_k, \mu_k)$, then

$$\sup_{\theta \in \Delta_{l,l}} |a_N(j, k; \theta) - a_N(j, k; \theta_l)|$$

$$\leq 8\pi\sqrt{2}[N^q]^{-1} \sum_{m=1}^{N} \sum_{n=1}^{N} |c(m - j, n - k)|\sqrt{m^2 + n^2}.$$

From these inequalities and the fact that $\sum_{m=1}^{N} \sum_{n=1}^{N} \sqrt{m^2 + n^2} = O(N^3)$ and the array $c(j, k)$ is absolutely summable, we obtain

$$E\left(\sup_{\theta \in \Delta_k, k} |S_N^*(\theta) - S_N^*(\theta_k)|\right) = O(N^{3-q}).$$

Since $q > 3$, the Markov Inequality and the Borel-Cantelli Lemma imply that

$$\sup_{\theta \in \Delta_k, k} |S_N^*(\theta) - S_N^*(\theta_k)| = o(N). \tag{4.7}$$

From (4.6),

$$P\left(\max_k |S_N^*(\theta_k)| > 2\pi\sqrt{2\beta\|f\|N^2 \ln(N^2)}\right)$$

$$\leq \sum_{k=1}^{K_N} P\left(|S_N^*(\theta_k)| > 2\pi\sqrt{2\beta\|f\|N^2 \ln(N^2)}\right)$$

$$\leq K_N \cdot 2N^{-2\alpha} = O(N^{2q-2\alpha}) \text{ for any } \beta > \alpha.$$

Since $2\alpha - 2q > 1$, the Borel-Cantelli Lemma implies that

$$\limsup_{N \to \infty} \frac{\max_k |S_N^*(\theta_k)|}{\sqrt{N^2 \ln(N^2)}} \leq 2\pi\sqrt{2\beta\|f\|}. \tag{4.8}$$

Since (4.8) is true for any $\beta > \alpha$, it is true for α. Inequality (4.5) now follows from (4.7) and (4.8).

Next, we show that $E\tilde{S}_N(\theta)$ is negligible. Since $E\varepsilon(j, k) = 0$,

$$E\tilde{S}_N(\theta) = -E \sum_j \sum_k a_N(j, k; \theta)\varepsilon(j, k)\mathbf{1}_{\{|a_N(j,k;\theta)\,\varepsilon(j,k)|\geq N/\ln(N^2)\}}.$$

Observing that

$$\sum_{j=-\infty}^{\infty} \sum_{k=-\infty}^{\infty} \sup_{\theta} |a_N(j, k; \theta)| \leq 2N^2 \sum_{j=-\infty}^{\infty} \sum_{k=-\infty}^{\infty} |c(j, k)|$$

and applying (4.4), gives

$$\sup_{\theta} |E\tilde{S}_N(\theta)| \leq \sum_{j} \sum_{k}$$

$$\times E\left(\sup_{\theta} |a_N(j,k;\theta)\varepsilon(j,k)| \mathbf{1}_{\{\sup_{\theta} |a_N(j,k;\theta)\,\varepsilon(j,k)| \geq N/\ln(N^2)\}}\right) \tag{4.9}$$

$$\leq \sum_{j} \sum_{k} \left(\frac{N}{\ln N^2}\right)^{1-r} E\left(\sup_{\theta} |a_N(j,k;\theta)| \, |\varepsilon(j,k)|\right)^{r}$$

$$\leq \left(\frac{N}{\ln N^2}\right)^{1-r} \sup_{j,k} E|\varepsilon(j,k)|^r \|a\|^{r-1} \sum_{j} \sum_{k} \sup_{\theta} |a_N(j,k;\theta)|$$

$$= O(N^{3-r}(\ln N)^{r-1}).$$

To complete the proof, it suffices to show that, with probability 1,

$$\sup_{\theta} |S_N^+(\theta) - \tilde{S}_N(\theta)| = o(N). \tag{4.10}$$

Since

$$\sup_{\theta} |S_N^+(\theta) - \tilde{S}_N(\theta)| \leq \sup_{\theta} \left| \sum_{j} \sum_{k} a_N(j,k;\theta)\varepsilon(j,k) \mathbf{1}_{\{|a_N(j,k;\theta)\,\varepsilon(j,k)| \geq N/\ln(N^2)\}} \right|,$$

$$E\left(\sup_{\theta} |S_N^+(\theta) - \tilde{S}_N(\theta)|\right) \leq$$

$$\sum_{j} \sum_{k} E\left(\sup_{\theta} |a_N(j,k;\theta)\varepsilon(j,k)| \mathbf{1}_{\{\sup_{\theta} |a_N(j,k;\theta)\,\varepsilon(j,k)| \geq N/\ln(N^2)\}}\right).$$

We see from the inequality below (4.9) that

$$E\left(\sup_{\theta} |S_N^+(\theta) - \tilde{S}_N(\theta)|\right) = O(N^{3-r}(\ln N)^{r-1}).$$

Since $r > 3$, the Borel-Cantelli Lemma implies (4.10). The proof is complete. $\square$

Let us now consider the periodogram for

$$y(m,n) = \sum_{k=1}^{p} \{C_k \cos(m\lambda_k + n\mu_k) + D_k \sin(m\lambda_k + n\mu_k)\} + x(m,n),$$

where $(m,n) \in \mathbb{Z}^2$, with $(\lambda, \mu) \in \Pi = (-\pi, \pi] \times [0, \pi]$, and $\mu_k \geq 0$ as stated before and $x(m,n)$ is of the form above as in Theorem 4.1.1. Then, we get that $I_N(\lambda, \mu; y)$ has a magnitude of order N^2 at any frequency (λ_j, μ_j), $j = 1, 2, \cdots, p$ and $O(\ln \ln N)$ elsewhere. That is,

Theorem 4.1.2. *If $\{y(m,n)\}$ and $\{x(m,n)\}$ are as above, then for any $(\lambda,\mu) \in \Pi$,*

$$I_N(\lambda,\mu;y) = \begin{cases} (A_j N/(4\pi))^2 + o(N^2) & \text{if } (\lambda,\mu) = (\lambda_j,\mu_j) \text{ for some } j \\ O(\ln\ln N) & \text{otherwise,} \end{cases}$$

where A_j, $j = 1,2,\cdots,p$ are constants.

Before we present the proof, we want to make some observations. Let $A_j > 0$, $\varphi_j \in [0,2\pi]$ such that $C_j = A_j\cos\varphi_j$, $D_j = A_j\sin\varphi_j$ for $j = 1,2,\cdots,p$ and extend λ_j, μ_j and φ_j to $j = -1,-2,\cdots,-p$, with $\lambda_j = -\lambda_{-j}$, $\mu_j = -\mu_{-j}$, and $\varphi_j = -\varphi_{-j}$. Define for $j = 0,\pm1,\pm2,\cdots,\pm p$, $B_0 = 0$, $B_j = A_{|j|}e^{i\varphi_j}$. We rewrite $y(m,n)$ as

$$y(m,n) = \frac{1}{2}\sum_{j=-p}^{p} B_j e^{i(\lambda_j m + \mu_j n)} + x(m,n).$$

Observe that

$$I_N(\lambda,\mu;y) = I_N(\lambda,\mu;x) + \frac{1}{(4\pi N)^2}\left|\sum_{k=-p}^{p} B_k H_N(\lambda_k - \lambda)H_N(\mu_k - \mu)\right|^2$$

$$+ \frac{1}{(2\pi N)^2}\,\mathrm{Re}\left(\sum_{m=1}^{N}\sum_{n=1}^{N} x(m,n)e^{-i(m\lambda+n\mu)}\sum_{k=-p}^{p} B_k H_N(\mu_k - \lambda)H_N(\lambda_k - \mu)\right),$$

$$(4.11)$$

where "Re" denotes the real part and $H_N(x) = \sum_{n=1}^{N} e^{inx}$. Finally, we recall that

$$|H_N(x)| = \begin{cases} N & \text{if } x = 0 \bmod 2\pi \\ \frac{|\sin(Nx/2)|}{|\sin(x/2)|} & \text{if } x \neq 0 \bmod 2\pi. \end{cases} \qquad (4.12)$$

Proof of Theorem 4.1.2: First, suppose that (λ,μ) is fixed, and not equal to (λ_j,μ_j) for all j. Then, from (4.12),

$$\left|\sum_{k=-p}^{p} B_k H_N(\lambda_k - \lambda)H_N(\mu_k - \mu)\right| = O(N),$$

which, together with (4.11) and Theorem 4.1.1, gives

$$I_N(\lambda,\mu;y) = O(\ln\ln N).$$

Now, suppose $(\lambda,\mu) = (\lambda_j,\mu_j)$, for some $j = 1,2,\cdots,p$. Then,

$$\frac{1}{N^2}\left|\sum_{k=-p}^{p} B_k H_N(\lambda_k - \lambda)H_N(\mu_k - \mu)\right| \longrightarrow |B_j| = A_j.$$

We see that (4.11) is dominated by the second term and

$$I_N(\lambda, \mu; y) = \left(\frac{A_j N}{4\pi}\right)^2 + o(N^2),$$

completing the proof. $\square$

The next theorem states the uniform behavior of $I_N(\lambda, \mu; y)$ in and out of neighborhoods of (λ_j, μ_j). For this we define for $\alpha > 0$, $j = 1, 2, \cdots, p$,

$$\Delta_{j,\alpha} = \{(\lambda, \mu) \in \Pi : |\lambda - \lambda_j| \le \pi/N^\alpha, |\mu - \mu_j| \le \pi/N^\alpha\}$$

and

$$\Delta_\alpha = \bigcup_{j=1}^p \Delta_{j,\alpha}.$$

With this notation, we now state the following theorem.

Theorem 4.1.3. *Suppose* $y(m, n) = z(m, n) + x(m, n)$ *for* $m, n \in \mathbb{Z}$ *and* $x(m, n)$ *is of the form described earlier with* $r > 3$. *Then, with probability 1, and for* N *sufficiently large,*

$$\inf_{(\lambda, \mu) \in \Delta_{j,1}} I_N(\lambda, \mu; y) > \left(\frac{A_j N}{\pi^3}\right)^2$$

and

$$\sup_{(\lambda, \mu) \in \Delta_\alpha^c} I_N(\lambda, \mu; y) = O(N^{2\alpha}),$$

where $\Delta_\alpha^c = \Pi \setminus \Delta_\alpha$.

Proof: Recall that $H_N(x) = \sum_{n=1}^N e^{inx}$. Then, $|H_N(x)|$ is symmetric on $(-\infty, \infty)$ and decreasing on $[0, \pi/N]$. Thus,

$$\inf_{|x| \le \pi/N} |H_N(x)| = |H_N(\pi/N)| = \frac{1}{\sin\left(\frac{\pi}{2N}\right)} > \frac{2N}{\pi}.$$

It follows that for any j,

$$\inf_{(\lambda, \mu) \in \Delta_{j,1}} |B_j H_N(\lambda_j - \lambda) H_N(\mu_j - \mu)| > \frac{4|B_j|N^2}{\pi^2}.$$

For $k \neq j$, we get from (4.12) that

$$\sup_{(\lambda, \mu) \in \Delta_{j,1}} |B_k H_N(\lambda_k - \lambda) H_N(\mu_k - \mu)| = O(N).$$

Hence, for sufficiently large N,

$$\inf_{(\lambda, \mu) \in \Delta_{j,1}} \left| \sum_{k=-p}^p B_k H_N(\lambda_k - \lambda) H_N(\mu_k - \mu) \right| > \frac{4|B_j|N^2}{\pi^2}.$$

Applying this, (4.11) and (4.12) give

$$
\inf_{(\lambda,\mu)\in\Delta_{j,1}} I_N(\lambda,\mu;y) \geq \inf_{(\lambda,\mu)\in\Delta_{j,1}} (4\pi N)^{-2} \left| \sum_{k=-p}^{p} B_k H_N(\lambda_k - \lambda) H_N(\mu_k - \mu) \right|^2
$$

$$
- \sup_{(\lambda,\mu)\in\Delta_{j,1}} (\pi N)^{-1} \sqrt{I_N(\lambda,\mu;x)} \left| \sum_{k=-p}^{p} B_k H_N(\lambda_k - \lambda) H_N(\mu_k - \mu) \right| > \frac{A_j^2 N^2}{\pi^6}.
$$

For $(\lambda,\mu) \in \Delta_\alpha^c$ and for any $j = 1,2,\cdots,p$, at least one of $|\lambda_j - \lambda|$, $|\mu_j - \mu|$ is greater than $\pi N^{-\alpha}$. Since

$$
\sup_{\pi N^{-\alpha} \leq |x| \leq 2\pi - \pi N^{-\alpha}} |H_N(x)| \leq \frac{1}{\sin\left(\frac{\pi}{2N^\alpha}\right)} \leq \frac{4N^\alpha}{\pi},
$$

we get

$$
\sup_{(\lambda,\mu)\in\Delta_\alpha^c} |H_N(\lambda - \lambda_j) H_N(\mu - \mu_j)| \leq N \frac{4N^\alpha}{\pi}.
$$

Consequently, $I_N(\lambda,\mu;y)$ is dominated by

$$
\frac{1}{(4\pi N)^2} \left(\sum_{j=-p}^{p} |B_j| N \frac{4N^\alpha}{\pi} \right)^2 = O(N^{2\alpha}). \quad \square
$$

From Theorem 4.1.2, we can see that $I_N(\lambda,\mu;y)$ has a magnitude of order N^2 at frequencies (λ_j,μ_j), $j = 1,2,\cdots,p$ and $O(\ln\ln N)$ elsewhere. Theorem 4.1.2 along with Theorem 4.1.3 provide ways to estimate p and hidden frequencies (λ_j,μ_j). We shall consider an approach which can be easily implemented on computers. Let α, β and c be constants such that

$$
0 < \alpha < 1, \quad 2\alpha < \beta < 2, \quad c > 0.
$$

Let $\Omega = \{(\lambda,\mu) \in \Pi : I_N(\lambda,\mu;y) > cN^\beta\}$. By Theorem 4.1.3, we get that for N sufficiently large Ω is contained in Δ_α and Ω contains some "clusters", which occur around (λ_j,μ_j). Let us define a cluster as a subset $\mathcal{S}$ of Ω such that

1. the diameter of $\mathcal{S}$ is no greater than $2\sqrt{2}\pi/N^\alpha$, and
2. for any $(\lambda,\mu) \in \Omega \setminus \mathcal{S}$ the diameter of $\{(\lambda,\mu)\} \cup \mathcal{S}$ is greater than $2\sqrt{2}\pi/N^\alpha$, where the diameter of $\mathcal{S}$ is given by

$$
\gamma(\mathcal{S}) = \sup\{\rho(x,y) : x \in \mathcal{S}, y \in \mathcal{S}\},
$$

where ρ is the Euclidean metric.

Let p_N be the number of clusters. We now prove that p_N is a consistent estimate of p, the number of frequencies.

Theorem 4.1.4. *Under the same conditions as in Theorem 4.1.3, with probability one, there will be exactly $p_N = p$ clusters for large N, each of which is of the form $\Delta_{j,\alpha} \cap \Omega$ and hence for any (λ, μ) in a cluster, $\rho((\lambda, \mu), (\lambda_j, \mu_j)) < 2\sqrt{2}\pi/N^\alpha$ for some j.*

Proof: It follows from Theorem 4.1.3 that if N is sufficiently large Ω is a subset of Δ_α and hence

$$\Omega = \bigcup_{j=1}^{p} (\Delta_{j,\alpha} \cap \Omega).$$

Let $S_j = \Delta_{j,\alpha} \cap \Omega$, $j = 1, 2, \ldots, p$. Then each S_j contains $\Delta_{j,1}$ and is non-empty. The diameter of S_j is bounded by that of $\Delta_{j,\alpha}$ which is $2\sqrt{2}\pi/N^\alpha$. For $(\lambda, \mu) \in \Omega \setminus S_j$, $(\lambda, \mu) \in S_k$ for some $k \neq j$. For $(\lambda', \mu') \in S_j$, using the Triangle Inequality twice, we get

$$\rho((\lambda, \mu), (\lambda', \mu')) \geq \rho((\lambda', \mu'), (\lambda_k, \mu_k)) - \rho((\lambda_k, \mu_k), (\lambda, \mu))$$

$$\geq \rho((\lambda_j, \mu_j), (\lambda_k, \mu_k)) - \rho((\lambda_j, \mu_j), (\lambda', \mu')) - \rho((\lambda_k, \mu_k), (\lambda, \mu)).$$

We see that the diameter of $\{(\lambda, \mu)\} \cap S_j$ is greater than or equal to

$$\rho((\lambda_j, \mu_j), (\lambda_k, \mu_k)) - 2\frac{2\sqrt{2}\pi}{N^\alpha}$$

which is greater than $2\sqrt{2}\pi/N^\alpha$ for N sufficiently large. Thus for large N, each S_j is a cluster by definition and there are exactly p clusters. For (λ, μ) in a cluster, say S_j, the distance between (λ, μ) and (λ_j, μ_j) is no longer than the diameter of S_j. This completes the proof. $\square$

We end this section by presenting an algorithm and giving some numerical results. For an integer $d > 0$, let

$$\omega_d = \left\{ \left(\frac{j\pi}{dN}, \frac{k\pi}{dN} \right) : j = 0, \pm 1, \ldots, \pm dN, k = 0, 1, \ldots, N \right\},$$

and

$$\Omega_d = \Omega \cap \omega_d.$$

Our algorithm consists of the following steps:

1. Calculate $I_N(\lambda, \mu; y)$ for $(\lambda, \mu) \in \omega_d$.

2. Identify the points in ω_d where $I_N(\lambda, \mu; y) > cN^\beta$.

3. Identify the clusters through the definition.

4. Also, find a value (λ, μ) that maximizes the periodogram in each of the clusters and take this frequency as an estimator of a hidden frequency.

The results earlier in this section are asymptotic, so the choices of the constants α, β, c and d do not affect the results when the sample size is sufficiently large. For a fixed sample size, the choices might be empirical. For example, we should choose a small c if one of the amplitudes is deemed small, but if the value of c is chosen too small, it may lead to overestimating p. α should be chosen closer to 1 to differentiate two frequencies which are close to each other. Simulation results can help us gain insights into the choices of α, β, c, and d for finite sample sizes. We used

$$\alpha = 3/4 \qquad \beta = 1.75 \qquad c = 6/\pi^6 \qquad d = 1 \text{ or } 4.$$

For the signal in model (4.1), we chose

$$p = 2 \qquad C_1 = D_1 = 1/\sqrt{2} \qquad C_2 = \sqrt{3} \qquad D_1 = 1$$

$$(\lambda_1, \mu_1) = (-0.1\pi, 0.3\pi) \qquad (\lambda_2, \mu_2) = (-0.5\pi, 0.7\pi),$$

and used four different noises:

1. i.i.d. standard normal
2. i.i.d. normal random field with mean 0 and standard deviation 2
3. A moving average

$$x(m, n) = 1/\sqrt{6} \sum_{j=-\infty}^{\infty} \sum_{k=0}^{\infty} 2^{-|j|-k} \varepsilon(m - j, n - k)$$

where $\varepsilon(m, n)$ are i.i.d. $N(0, 1)$. We denote this noise by MA_1.

4. A moving average

$$x(m, n) = 2/\sqrt{6} \sum_{j=-\infty}^{\infty} \sum_{k=0}^{\infty} 2^{-|j|-k} \varepsilon(m - j, n - k)$$

where $\varepsilon(m, n)$ are i.i.d. $N(0, 1)$. We denote this noise by MA_2.

Then,

$$E|x(m, n)|^2 = \begin{cases} 1 & \text{for the } MA_1 \text{ noise} \\ 4 & \text{for the } MA_2 \text{ noise.} \end{cases}$$

Therefore, we can compare the estimators from the colored noises to those that are from the white noises. For each of the four noises, we simulated observations $y(m, n)$, for $m = 1, 2, \ldots, 19$ to evaluate the performance of the estimation for small samples. We expect larger sample sizes only yield better estimators with the same choice of α, β, c, and d.

The points where $I_N(\lambda, \mu; y) > cN^\beta$ naturally occur in "clustered sets". However, we need to check that each such "clustered set" has a diameter no larger than $2\sqrt{2}\pi/N^\alpha$, and adding an additional point will increase the diameter to more than $2\sqrt{2}\pi/N^\alpha$.

When we used $d = 1$, p is correctly estimated 100% of the time for i.i.d. $N(0, 1)$ and MA_1 noises, and overestimated once ($\hat{p} = 3$) for the i.i.d. $N(0, 4)$ noise, and overestimated twice ($\hat{p} = 3$) and underestimated once ($\hat{p} = 1$) for the MA_2 noise. The average of the estimated frequencies when p is correctly estimated is shown in Table 4.1. We see that a higher signal-noise ratio (hence a smaller variance of the noise) generally improves the estimation of p, but does not greatly affect the precision of the estimation of frequencies. Since the colored noise does not have a special density as flat as that of a white noise, estimation of p may require larger sample sizes in the presence of colored noise. However, as seen from Table 4.1, colored noises do not greatly affect the estimation of the frequencies if p is estimated correctly. Since the estimators for the frequencies may be biased, the mean squared errors are reported.

Similar conclusions can be made with $d = 4$. When d is increased to 4, estimation for p and the frequencies generally become better. The number p is correctly estimated 100% of the time for all noises except the MA_2 noise, for which p is correctly estimated 94 times and overestimated 6 times. The average of the estimated frequencies when p is correctly estimated is shown in Table 4.2.

We note that $N = 19$ is a small number. Even so, our choices of α, β, c, and d yield early satisfactory results. We expect the estimation for p and the frequencies to become better with a large N since the peak near a hidden frequency will become stronger. In fact, we used $N = 39$ for the MA_2 noise and $d = 4$, which gave the worst estimation results for $N = 19$. Again, we ran 100 simulations. The number p was correctly estimated 100 times and the averages of estimated frequencies and the mean squared errors are provided in Table 4.3. The results become better.

TABLE 4.1

Averages of Estimates of Frequencies (in Multiples of π) with $d = 1$.

Noise	λ_1/π	μ_1/π	λ_2/π	μ_2/π
i.i.d. $N(0,1)$	-0.1052632	0.3157895	-0.4978947	0.6842105
	(0.0000277)	(0.0002493)	(0.0006925)	(0.0002493)
i.i.d. $N(0,4)$	-0.1052632	0.3135965	-0.5010965	0.6842105
	(0.0000277)	(0.0002955)	(0.0006925)	(0.0002493)
MA_1	-0.1052632	0.3157895	-0.4821053	0.6842105
	(0.0000277)	(0.0002493)	(0.0006925)	(0.0002493)
MA_2	-0.1030235	0.3101904	-0.4837626	0.6842105
	(0.0001220)	(0.0003672)	(0.0006925)	(0.0002493)

Note: Values in parentheses are the mean squared error of the estimates. Results are based on 100 simulations.

TABLE 4.2

Averages of Estimates of Frequencies (in Multiples of π) with $d = 4$.

Noise	λ_1/π	μ_1/π	λ_2/π	μ_2/π
i.i.d. N(0,1)	−0.0994737	0.3018421	−0.5	0.6981579
	(0.0000429)	(0.0000132)	(0.0)	(0.0000132)
i.i.d. N(0,4)	−0.0993555	0.3018260	−0.5	0.6981740
	(0.0000432)	(0.0000133)	(0.0)	(0.0000133)
MA_1	−0.0993555	0.3018260	−0.5	0.6981740
	(0.0000432)	(0.0000133)	(0.0)	(0.0000133)
MA_2	−0.0983852	0.3002392	−0.5	0.6973684
	(0.0001166)	(0.0000573)	(0.0)	(0.0000069)

Note: Values in parentheses are the mean squared error of the estimates. Results are based on 100 simulations.

TABLE 4.3

Averages of Estimates of Frequencies (in Multiples of π) with $d = 4$, $N = 39$.

Noise	λ_1/π	μ_1/π	λ_2/π	μ_2/π
MA_2	−0.1003205	0.03	−0.5	0.6987179
	(0.0000095)	(0.0000066)	(0.0)	(0.0000016)

Note: Values in parentheses are the mean squared error of the estimates. Results are based on 100 simulations.

4.2 Invariant Subspaces of $L^p(\mathbb{T})$

We will write $L^p(\mathbb{T})$, for $1 < p < \infty$, to denote the collection of all Lebesgue measurable functions $f : \mathbb{T} \to \mathbb{C}$ such that

$$\int_{[-\pi,\pi)} |f(e^{i\lambda})|^p \, d\sigma(e^{i\lambda}) < \infty.$$

As always, we will identify functions that are equal $[\sigma]$-a.e., where σ denotes normalized Lebesgue measure on $\mathbb{T}$. An important subspace of $L^p(\mathbb{T})$ that will come into play is

$$H^p(\mathbb{T}) = \overline{span} \left\{ e^{im\lambda} : m \geq 0 \right\}.$$

Let $S : L^p(\mathbb{T}) \to L^p(\mathbb{T})$ be the linear operator defined by $S(f)(e^{i\lambda}) = e^{i\lambda} f(e^{i\lambda})$. It is straightforward to verify that S is an isometry in $\mathcal{L}(L^p(\mathbb{T}))$. A subspace $\mathcal{M}$ of $L^p(\mathbb{T})$ is said to be S-invariant if $S(\mathcal{M}) \subseteq \mathcal{M}$. Further, we say that $\mathcal{M}$ is S-simply invariant if $S(\mathcal{M})$ is a proper subspace of $\mathcal{M}$ and S-doubly invariant if $S(\mathcal{M}) = \mathcal{M}$.

Our approach in this section will be slightly different than that used in most books. We will follow the approach used in [32], which uses some facts from approximation theory. Recall that $L^p(\mathbb{T})$ and all of its subspaces are uniformly convex Banach spaces. Hence, we may employ the following lemmas found in [38].

Lemma 4.2.1. *Let $\mathcal{X}$ be a uniformly convex Banach space and let $\mathcal{K}$ be a subspace of $\mathcal{X}$. Then, for all x in $\mathcal{X}$ there corresponds a unique y in $\mathcal{K}$ satisfying $\|x - y\| = \inf_{z \in \mathcal{K}} \|x - z\|$. We call such y the best approximate of x in $\mathcal{K}$ and denote it by x^*.*

Next, we introduce the concept of orthogonality in a uniformly convex Banach space. In a uniformly convex Banach space $\mathcal{X}$, we say w in $\mathcal{X}$ is orthogonal to $\mathcal{K}$, and write $w \perp \mathcal{K}$, if $\|w\| \le \|w+k\|$ for all k in $\mathcal{K}$. Note that if x in any element of $\mathcal{X}$ and x^* is the best approximate of x in $\mathcal{K}$, then $(x - x^*)$ is orthogonal to $\mathcal{K}$. From this observation, we get the following lemma.

Lemma 4.2.2. *Let $\mathcal{X}$ be a uniformly convex Banach space and let $\mathcal{K}$ be a subspace of $\mathcal{X}$. Then, there exists an x in $\mathcal{X}$ such that $x \perp \mathcal{K}$. If $\mathcal{K}$ is a proper subspace of $\mathcal{X}$, then x can be chosen such that $x \neq 0$.*

The following lemma gives us an alternative way to check orthogonality in $L^p(\mathbb{T})$.

Lemma 4.2.3. *Let $\mathcal{K}$ be a closed subspace of $L^p(\mathbb{T})$. Then, an element $f \in L^p(\mathbb{T})$ is orthogonal to $\mathcal{K}$ if and only if*

$$\int_{[-\pi,\pi)} |f(e^{i\lambda})|^{p-1} \, \overline{\mathrm{sgn}(f(e^{i\lambda}))} \cdot k(e^{i\lambda}) \, d\sigma(e^{i\lambda}) = 0$$

for all k in $\mathcal{K}$.

Here, $\mathrm{sgn}(f(e^{i\lambda}))$ is a complex measurable function of modulus one such that $f(e^{i\lambda}) = |f(e^{i\lambda})|\mathrm{sgn}(f(e^{i\lambda}))$.

We are now ready to prove our main results.

Theorem 4.2.1. *Every S-doubly invariant subspace of $L^p(\mathbb{T})$ is of the form $1_E L^p(\mathbb{T})$, where E is a measurable subset of $\mathbb{T}$ and 1_E is the indicator function of E.*

Proof: We will leave it to the reader to verify that every subspace of the form $1_E L^p(\mathbb{T})$, where E is a measurable subset of $\mathbb{T}$, is an S-doubly invariant subspace of $L^p(\mathbb{T})$. We will prove the other direction.

Let $\mathcal{M}$ denote the S-doubly invariant subspace in question. If $1_{\mathbb{T}}$ is in $\mathcal{M}$, then $\mathcal{M} = L^p(\mathbb{T})$ and the theorem in proved. If, however, $1_{\mathbb{T}}$ is not in $\mathcal{M}$, then let q denote the best approximate of $1_{\mathbb{T}}$ in $\mathcal{M}$. Therefore, $1_{\mathbb{T}} - q$ is orthogonal to $\mathcal{M}$. Hence, $(1_{\mathbb{T}} - q) \perp S^m q$ for all $m \in \mathbb{Z}$. That is,

$$\int_{[-\pi,\pi)} |(1_{\mathbb{T}} - q)(e^{i\lambda})|^{p-1} \, \overline{\mathrm{sgn}((1_{\mathbb{T}} - q)(e^{i\lambda}))} \cdot e^{im\lambda} q(e^{i\lambda}) \, d\sigma(e^{i\lambda}) = 0$$

for all $m \in \mathbb{Z}$. Therefore, $\left|(\mathbf{1}_{\mathbb{T}} - q)(e^{i\lambda})\right|^{p-1} \overline{\operatorname{sgn}((\mathbf{1}_{\mathbb{T}} - q)(e^{i\lambda}))} \cdot q(e^{i\lambda}) = 0$
$[\sigma]$-a.e. It follows from this that q takes only values zero and one $[\sigma]$-a.e. If we
let $E = \{e^{i\lambda} : q(e^{i\lambda}) = 1\}$, it follows the $q = \mathbf{1}_E$ $[\sigma]$-a.e. Since $\mathbf{1}_E L^p(\mathbb{T})$ is the
smallest S-doubly invariant subspace containing $\mathbf{1}_E$, it follows that $\mathbf{1}_E L^p(\mathbb{T})$
is a subspace of $\mathcal{M}$. To see that $\mathbf{1}_E L^p(\mathbb{T})$ is all of $\mathcal{M}$, let g be an element of
$\mathcal{M}$ orthogonal to $\mathbf{1}_E L^p(\mathbb{T})$. So, in particular, $g \perp S^m \mathbf{1}_E$ for all $m \in \mathbb{Z}$. That
is,

$$\int_{[-\pi,\pi)} \left|g(e^{i\lambda})\right|^{p-1} \overline{\operatorname{sgn}(g(e^{i\lambda}))} \cdot e^{im\lambda} \mathbf{1}_E(e^{i\lambda}) \, d\sigma(e^{i\lambda}) = 0$$

for all $m \in \mathbb{Z}$. Therefore, $\left|g(e^{i\lambda})\right|^{p-1} \overline{\operatorname{sgn}(g(e^{i\lambda}))} \cdot \mathbf{1}_E(e^{i\lambda}) = 0$ $[\sigma]$-a.e. It follows
from this that g equals zero on E. It remains to show that g equals zero on
E^c. Since g is in $\mathcal{M}$, it follows that $S^m g$ is in $\mathcal{M}$ for all $m \in \mathbb{Z}$. Therefore,
$\mathbf{1}_{E^c} = (\mathbf{1}_{\mathbb{T}} - \mathbf{1}_E) \perp S^m g$ for all $m \in \mathbb{Z}$. That is,

$$\int_{[-\pi,\pi)} \left|\mathbf{1}_{E^c}(e^{i\lambda})\right|^{p-1} \overline{\operatorname{sgn}(\mathbf{1}_{E^c}(e^{i\lambda}))} \cdot e^{im\lambda} g(e^{i\lambda}) \, d\sigma(e^{i\lambda}) = 0$$

for all $m \in \mathbb{Z}$. Therefore, $\left|\mathbf{1}_{E^c}(e^{i\lambda})\right|^{p-1} \overline{\operatorname{sgn}(\mathbf{1}_{E^c}(e^{i\lambda}))} \cdot g(e^{i\lambda}) = 0$ $[\sigma]$-a.e.
Therefore, g must also equal zero on E^c. It then follows that $\mathcal{M} = \mathbf{1}_E L^p(\mathbb{T})$
as desired. $\square$

Theorem 4.2.2. *$\mathcal{M}$ is an S-simply invariant subspace of $L^p(\mathbb{T})$ if and only
if $\mathcal{M} = \phi H^p(\mathbb{T})$ with ϕ unimodular.*

Proof: If $\mathcal{M} = \phi H^p(\mathbb{T})$ with ϕ unimodular, then it is clear that $\mathcal{M}$ is an
S-simply invariant subspace of $L^p(\mathbb{T})$. It remains to show the converse. Since
$S(\mathcal{M})$ is a proper subspace of $\mathcal{M}$ by Lemma 4.2.2 there exists a nonzero ϕ in
$\mathcal{M}$ such that $\phi \perp S(\mathcal{M})$. There is no loss of generality if we choose ϕ such
that $\|\phi\|_p = 1$. So in particular, $\phi \perp S^n \phi$ for all $n > 0$. That is,

$$\int_{[-\pi,\pi)} e^{in\lambda} |\phi(e^{i\lambda})|^p \, d\sigma(e^{i\lambda}) = 0$$

for all $n > 0$. Taking complex conjugates we get

$$\int_{[-\pi,\pi)} e^{in\lambda} |\phi(e^{i\lambda})|^p \, d\sigma(e^{i\lambda}) = 0$$

for all $n \neq 0$. So, $|\phi| = 1$ a.e. on $\mathbb{T}$. That is, ϕ is unimodular. Since ϕ is in
$\mathcal{M}$, so is $e^{in\lambda}\phi(e^{i\lambda})$ for all $n \geq 0$. Therefore, $\phi(e^{i\lambda})P(e^{i\lambda})$ is in $\mathcal{M}$ for every
polynomial P. Since polynomials are dense in $H^p(\mathbb{T})$ and $|\phi| = 1$, we get that
$\phi H^p(\mathbb{T}) \subseteq \mathcal{M}$. It remains to show that $\phi H^p(\mathbb{T})$ is all of $\mathcal{M}$. Let ψ be an
element of $\mathcal{M}$ orthogonal to $\phi H^p(\mathbb{T})$. Since ϕ is unimodular, we get that $\psi\overline{\phi}$
is in $L^p(\mathbb{T})$. By the way we chose ϕ we get that $\phi \perp S^n \psi$ for all $n > 0$. That
is,

$$\int_{[-\pi,\pi)} e^{in\lambda} \psi(e^{i\lambda})\overline{\phi}(e^{i\lambda}) \, d\sigma(e^{i\lambda}) = 0$$

for all $n > 0$. These two facts together give us that $\psi\bar{\phi}$ is in $H^p(\mathbb{T})$. That is, ψ is in $\phi H^p(\mathbb{T})$. This can only happen if $\psi = 0$. Therefore, $\mathcal{M} = \phi H^p(\mathbb{T})$ as desired. $\square$

Corollary 4.2.1. $\mathcal{M}$ *is an* S-*invariant subspace of* $H^p(\mathbb{T})$ *if and only if* $\mathcal{M} = \phi H^p(\mathbb{T})$ *with* ϕ *inner.*

This is easy to see since every S-invariant subspace of $H^p(\mathbb{T})$ is S-simply invariant and unimodular plus analytic implies inner.

4.3   Harmonizable $S\alpha S$ Sequences

In recent years, based on data from finance, insurance and hydrology, it is found that one needs to study the stationary random sequences that are not second order $\left(\text{i.e., } E|X_n|^2 = \infty\right)$. For this, one needs to study the models given by so-called stable sequences. In this section, we follow the development of this material given by S. Cambanis and A. Miamee in [3].

A real-valued random variable X is called symmetric α-stable, $0 < \alpha \le 2$, abbreviated $S\alpha S$, if its characteristic function $\varphi_X(t) = E\exp\{itX\}$, $t \in \mathbb{R}$ has the form

$$\varphi_X(t) = \exp\{-c|t|^\alpha\}, \quad t \in \mathbb{R},$$

for some $c \ge 0$. A finite collection of real-valued random variables are called jointly $S\alpha S$ if all linear combinations of these random variables are $S\alpha S$. A complex random variable $Z = X + iY$ is called isotropic α-stable if X and Y are jointly $S\alpha S$ and Z has a radially symmetric distribution; that is, $e^{i\theta}Z$ and Z have the same distribution for all real θ. This is equivalent to the following requirement on $\varphi_Z(t) = E\exp\{i\operatorname{Re}(\bar{t}Z)\}$, $t \in \mathbb{C}$, the characteristic function of Z,

$$\varphi_Z(t) = \exp\{-c|t|^\alpha\}, \quad t \in \mathbb{C},$$

for some $c \ge 0$. As in [15], we can define a length on Z, by

$$\|Z\| = \begin{cases} c^{1/\alpha}, & \text{for } 1 \le \alpha \le 2 \\ c, & \text{for } 0 < \alpha < 1. \end{cases}$$

This length gives a metric on any family of random variables with the property that any linear combination of its members is an isotropic α-stable random variable. We point out that if Z_1 and Z_2 are independent, then

$$\|Z_1 + Z_2\|^\alpha = \|Z_1\|^\alpha + \|Z_2\|^\alpha \quad \text{for } 1 \le \alpha \le 2$$

and

$$\|Z_1 + Z_2\| = \|Z_1\| + \|Z_2\| \quad \text{for } 0 < \alpha < 1.$$

Now, let $\mathcal{Z}$ be an independently scattered complex isotropic α-stable variable valued set function defined on $\mathcal{B}(\mathbb{T})$, the Borel subsets of $\mathbb{T}$. That is, for all disjoint sets $\Delta_1, \cdots, \Delta_n \in \mathcal{B}(\mathbb{T})$, $\mathcal{Z}(\Delta_1), \cdots, \mathcal{Z}(\Delta_n)$ are independent with

$$\varphi_{\mathcal{Z}(\Delta_k)}(t) = \begin{cases} \exp\left\{-|t|^\alpha \|\mathcal{Z}(\Delta_k)\|^\alpha\right\} & \text{for } 1 \le \alpha \le 2 \\ \exp\left\{-|t|^\alpha \|\mathcal{Z}(\Delta_k)\|\right\} & \text{for } 0 < \alpha < 1 \end{cases}, \quad t \in \mathbb{C}.$$

Using $\mathcal{Z}$, we define

$$\mu(\Delta) = \begin{cases} \|\mathcal{Z}(\Delta)\|^\alpha & \text{for } 1 \le \alpha \le 2 \\ \|\mathcal{Z}(\Delta)\| & \text{for } 0 < \alpha < 1 \end{cases}, \quad \text{for } \Delta \in \mathcal{B}(\mathbb{T}),$$

and observe that μ is a finite measure defined on $\mathcal{B}(\mathbb{T})$. From this, it follows that if $f \in L^\alpha(\mathbb{T}, \mu)$ and $\mathcal{X} = \int_{[-\pi,\pi)} f(e^{i\lambda}) \, d\mathcal{Z}(e^{i\lambda})$, then

$$\varphi_{\mathcal{X}}(t) = \exp\left\{-|t|^\alpha \int_{[-\pi,\pi)} \left|f(e^{i\lambda})\right|^\alpha \, d\mu(e^{i\lambda})\right\}, \quad t \in \mathbb{C}.$$

A complex random sequence X_n, $n \in \mathbb{Z}$ is called harmonizable $S\alpha S$ with spectral measure μ, if μ is a finite (positive) measure defined on $\mathcal{B}(\mathbb{T})$ with

$$E \exp\left\{i \operatorname{Re}\left(\bar{t} \sum_{j=1}^N z_j X_{t_j}\right)\right\} = \exp\left\{-|t|^\alpha \int_{[\pi,\pi)} \left|\sum_{j=1}^N z_j e^{-it_j \lambda}\right|^\alpha \, d\mu(e^{i\lambda})\right\},$$

where $t, z_j \in \mathbb{C}$, and $t_j \in \mathbb{Z}$, for $j = 1, \cdots, N$. We see from this that X_n, $n \in \mathbb{Z}$ is (strictly) stationary. We may define a harmonizable $S\alpha S$ sequence equivalently through its spectral representation

$$X_n = \int_{[-\pi,\pi)} e^{-in\lambda} \, d\mathcal{Z}(e^{i\lambda}),$$

where $\mathcal{Z}$ is an independently scattered complex isotropic α-stable variable valued set function defined on $\mathcal{B}(\mathbb{T})$. If $L(X)$ is the closure in probability of the linear span of X_n, $n \in \mathbb{Z}$, then the correspondence between f and $\int_{[-\pi,\pi)} f(e^{i\lambda}) \, d\mathcal{Z}(e^{i\lambda})$ gives an isomorphism between $L^\alpha(\mathbb{T}, \mu)$ and $L(X)$, that sends $e^{-in\lambda}$ to X_n. Hence, every $Y \in L(X)$ has a representation of the form $\int_{[-\pi,\pi)} f(e^{i\lambda}) \, d\mathcal{Z}(e^{i\lambda})$ for some $f \in L^\alpha(\mathbb{T}, \mu)$ and has a radially symmetric distribution.

When $1 < \alpha \leq 2$ and $Y_1, Y_2 \in L(X)$ are represented by $f_1, f_2 \in L^\alpha(\mathbb{T}, \mu)$, the covariation of Y_1 with Y_2 is defined by

$$[Y_1, Y_2]_\alpha = \int_{[-\pi,\pi)} f_1(e^{i\lambda}) |f_2(e^{i\lambda})|^{\alpha-1} \overline{\mathrm{sgn}(f_2(e^{i\lambda}))} \, d\mu(e^{i\lambda}),$$

where $\mathrm{sgn}(f(e^{i\lambda}))$ is a complex measurable function of modulus one such that $f(e^{i\lambda}) = |f(e^{i\lambda})|\mathrm{sgn}(f(e^{i\lambda}))$. By Hölder's inequality, we have that $|[Y_1, Y_2]_\alpha| \leq \|f_1\|_{L^\alpha(\mathbb{T},\mu)} \|f_2\|_{L^\alpha(\mathbb{T},\mu)}^{\alpha-1}$ with equality if and only if $Y_1 = zY_2$, where $z \in \mathbb{C}$. The covariation of the harmonizable $S\alpha S$ sequence has the form

$$[X_n, X_m]_\alpha = \int_{[-\pi,\pi)} e^{-i(n-m)\lambda} \, d\mu(e^{i\lambda}).$$

Note that this is the form of the covariance of a weakly stationary sequence.

For $Y_1, Y_2 \in L(X)$, if $[Y_1, Y_2]_\alpha = 0$, we say that Y_2 is orthogonal to Y_1, and write $Y_2 \perp Y_1$, which is nonsymmetric and was introduced by R. C. James in [17]. When $Y_2 \perp Y_1$ and $Y_1 \perp Y_2$, we say that Y_1 and Y_2 are mutually orthogonal. We now make an important distinction between the Gaussian case, $\alpha = 2$, and the non-Gaussian case, $1 < \alpha < 2$, when it comes to the relationship between independence and orthogonality. When $\alpha = 2$, the independence of Y_1 and Y_2 is equivalent to the mutual orthogonality of Y_1 and Y_2 and the mutual orthogonality of Y_1 and $\overline{Y_2}$. However, when $1 < \alpha < 2$, the independence of Y_1 and Y_2 implies the mutual orthogonality of Y_1 and Y_2 and the mutual orthogonality of Y_1 and $\overline{Y_2}$, but it is **not generally true** that the mutual orthogonality of Y_1 and Y_2 and the mutual orthogonality of Y_1 and $\overline{Y_2}$ implies the independence of Y_1 and Y_2. This is because when $1 < \alpha < 2$, the independence of Y_1 and Y_2 is equivalent to their representing function f_1 and f_2 having disjoint support; that is, $f_1 \cdot f_2 = 0$ $[\mu]$-a.e., see [40], while mutual orthogonality of Y_1 and Y_2 means, by definition, that $\int_{[-\pi,\pi)} f_1(e^{i\lambda}) |f_2(e^{i\lambda})|^{\alpha-1} \overline{\mathrm{sgn}(f_2(e^{i\lambda}))} \, d\mu(e^{i\lambda}) =$

$$0 = \int_{[-\pi,\pi)} f_2(e^{i\lambda}) |f_1(e^{i\lambda})|^{\alpha-1} \overline{\mathrm{sgn}(f_1(e^{i\lambda}))} \, d\mu(e^{i\lambda}).$$

For a harmonizable $S\alpha S$ sequence X_n, $n \in \mathbb{Z}$, we define the concept of regularity and singularity. Let $L(X : n)$ denote the closure in probability of the linear span of $\{X_k : k \leq n\}$ and let $L(X : -\infty) = \cap_n L(X : n)$. X_n, $n \in \mathbb{Z}$ is called regular if $L(X : -\infty) = \{0\}$ and singular if $L(X : -\infty) = L(X)$. These definitions are consistent with those given for weakly stationary sequences.

Theorem 4.3.1. *Let X_n, $n \in \mathbb{Z}$ be a harmonizable $S\alpha S$ sequence with $1 < \alpha \leq 2$ and spectral measure μ. Then, the following are equivalent.*

1. X_n, $n \in \mathbb{Z}$ is regular.

2. μ is absolutely continuous with respect to σ with density $f \geq 0$ $[\sigma]$-a.e. satisfying

$$\int_{[-\pi,\pi)} \log(f(e^{i\lambda})) \, d\sigma(e^{i\lambda}) > -\infty.$$

3. μ is absolutely continuous with respect to σ with density $f(e^{i\lambda}) = |\varphi(e^{i\lambda})|^{\alpha}$, where φ is an outer function in $H^{\alpha}(\mathbb{T})$.

4. X_n, $n \in \mathbb{Z}$ has a unique moving average representation

$$X_n = \sum_{k=0}^{\infty} a_k V_{n-k},$$

where $a_0 > 0$ and the random sequence V_n, $n \in \mathbb{Z}$ is jointly stationary with X_n, $n \in \mathbb{Z}$, satisfies $L(X : n) = L(V : n)$, and is a harmonizable $S\alpha S$ sequence with spectral measure σ and thus consists of mutually orthogonal random variables with norm one.

As usual, $H^{\alpha}(\mathbb{T})$ is the closure in $L^{\alpha}(\mathbb{T}, \sigma)$ of the linear span of $\{e^{ik\lambda} : k \geq 0\}$. Recall that outer functions are unique up to a constant multiple of modulus one. For this reason, as you will see in the proof, the a_k's are, up to a constant multiple of modulus one, the Fourier coefficients of the outer function φ.

Proof. $(1) \Rightarrow (2)$: Suppose X_n, $n \in \mathbb{Z}$ is regular. By definition, $L(X : -\infty) = \{0\}$. Using the isomorphism that takes X_n to $e^{in\lambda}$ in $L^{\alpha}(\mathbb{T}, \mu)$, defining $M_n^{(\alpha)} = \overline{span}\{e^{ik\lambda} : k \leq n\}$ and $M_{-\infty}^{(\alpha)} = \cap_n M_n^{(\alpha)}$, it follows that if X_n is regular, then $M_{-\infty}^{(\alpha)} = \{0\}$. Since $1 < \alpha \leq 2$, it follows that $M_n^{(2)} \subseteq M_n^{(\alpha)}$, for $n \in \mathbb{Z}$, and hence $M_{-\infty}^{(2)} \subseteq M_{-\infty}^{(\alpha)} = \{0\}$. It then follows from Corollary 1.15.2 that μ is absolutely continuous with respect to σ with density $f \geq 0$ $[\sigma]$-a.e. satisfying $\int_{[-\pi,\pi)} \log(f(e^{i\lambda})) \, d\sigma(e^{i\lambda}) > -\infty$.

$(2) \Rightarrow (3)$: Suppose that μ is absolutely continuous with respect to σ with density $f \geq 0$ $[\sigma]$-a.e. satisfying $\int_{[-\pi,\pi)} \log(f(e^{i\lambda})) \, d\sigma(e^{i\lambda}) > -\infty$. It follows that

$$\int_{[-\pi,\pi)} \log(f^{\frac{2}{\alpha}}(e^{i\lambda})) \, d\sigma(e^{i\lambda}) = \frac{2}{\alpha} \int_{[-\pi,\pi)} \log(f(e^{i\lambda})) \, d\sigma(e^{i\lambda}) > -\infty.$$

By the first theorem on page 53 of [13], it follows that $f^{\frac{2}{\alpha}} = |\varphi|^2$ for some $\varphi \in H^2(\mathbb{T}) \subset H^{\alpha}(\mathbb{T})$. Without loss of generality, we may take φ to be outer. Therefore, $f(e^{i\lambda}) = |\varphi(e^{i\lambda})|^{\alpha}$, where φ is an outer function in $H^{\alpha}(\mathbb{T})$.

$(3) \Rightarrow (4)$: Suppose μ is absolutely continuous with respect to σ with density $f(e^{i\lambda}) = |\varphi(e^{i\lambda})|^{\alpha}$, where φ is an outer function in $H^{\alpha}(\mathbb{T})$. Without loss of generality, we can assume that $\hat{\varphi}(0) > 0$. Indeed, if it were not, we can multiply φ by a constant of modulus one, thus making $\hat{\varphi}(0) > 0$ and keeping φ outer. Now, let $U_1 : L^\alpha(\mathbb{T}, f \, d\sigma) \to L(X)$ be defined by $U_1(g) = \int_{[-\pi,\pi)} g(e^{i\lambda}) \, d\mathcal{Z}(e^{i\lambda})$. This is a linear isometry that is onto. Also, let $U_2 : L^\alpha(\mathbb{T}, f \, d\sigma) \to L^\alpha(\mathbb{T}, \sigma)$ be defined by $U_2(g) = g\varphi$. This is a linear isometry and is also onto since φ is outer. Then, $U = U_2 U_1^{-1} : L(X) \to L^\alpha(\mathbb{T}, \sigma)$ and $U(X_n) = U_2 U_1^{-1}(X_n) = U_2(e^{-in\lambda}) = e^{-in\lambda}\varphi(e^{i\lambda})$. Let $V_n = U^{-1}(e^{-in\lambda})$ and the fact that φ is outer gives us that $L(X : n) = L(V : n)$, $n \in \mathbb{Z}$. Now, since $\varphi \in H^\alpha(\mathbb{T})$, it has a Fourier series

$$\varphi(e^{i\lambda}) = \sum_{k=0}^{\infty} a_k e^{ik\lambda},$$

which converges in $L^\alpha(\mathbb{T}, \sigma)$. Recall that $a_0 = \hat{\varphi}(0) > 0$. Therefore,

$$
\begin{aligned}
X_n &= U^{-1}(e^{-in\lambda}\varphi(e^{i\lambda})) = U^{-1}\left(e^{-in\lambda}\sum_{k=0}^{\infty} a_k e^{ik\lambda}\right) \\
&= U^{-1}\left(\sum_{k=0}^{\infty} a_k e^{-i(n-k)\lambda}\right) = \sum_{k=0}^{\infty} a_k V_{n-k} \quad \text{in } L(X).
\end{aligned}
$$

In view of our isomorphism U, we have that

$$E \exp\left\{ i \operatorname{Re}\left(\bar{t} \sum_{j=1}^{N} z_j V_{t_j}\right)\right\} = \exp\left\{ -|t|^\alpha \int_{[\pi,\pi)} \left|\sum_{j=1}^{N} z_j e^{-it_j\lambda}\right|^\alpha d\sigma(e^{i\lambda})\right\},$$

where $t, z_j \in \mathbb{C}$, and $t_j \in \mathbb{Z}$, for $j = 1, \cdots, N$. Thus, V_n, $n \in \mathbb{Z}$ is harmonizable SαS with spectral measure σ and thus

$$[V_k, V_n]_\alpha = \int_{[-\pi,\pi)} e^{-ik\lambda} e^{in\lambda} \, d\sigma(e^{i\lambda}) = \delta_{kn},$$

and so the V_n's are mutually orthogonal with $\|V_n\|_\alpha = [V_n, V_n]_\alpha^{1/\alpha} = 1$, for all $n \in \mathbb{Z}$. The joint stationarity of X_n, $n \in \mathbb{Z}$ and V_n, $n \in \mathbb{Z}$ can be seen from the fact that $X_n = U^{-1}(e^{-in\lambda}\varphi(e^{i\lambda}))$ and $V_n = U^{-1}(e^{-in\lambda})$. To see that this representation is unique, suppose that there is another such representation $X_n = \sum_{k=0}^{\infty} b_k W_{n-k}$ satisfying the conditions given in (4). By the orthogonality of the V_n's and W_n's, it follows that $X_n - X_n^* = a_0 V_n = b_0 W_n$, where X_n^* is the best approximate of X_n in $L(X : n-1) = L(V : n-1) = L(W : n-1)$. Now, since $\|V_n\|_\alpha = \|W_n\|_\alpha = 1$ for all $n \in \mathbb{Z}$, it follows that $|a_0| = |b_0|$, and since a_0 and b_0 are positive, it follows that $a_0 = b_0$. Therefore, $V_n = W_n$ for all $n \in \mathbb{Z}$. It then follows that $\sum_{k=0}^{\infty}(a_k - b_k)V_{n-k} = 0$. This together with the orthogonality of the V_n's give us that $a_n = b_n$ for all $n \in \mathbb{Z}$.

(4) $\Rightarrow$ (1): Suppose that $X_n = \sum_{k=0}^{\infty} a_k V_{n-k}$, where all the conditions of

(4) hold. Since $L(X : n) = L(V : n)$, $n \in \mathbb{Z}$, it follows that $L(X : -\infty) = L(V : -\infty)$. We will show that $L(V : -\infty) = \{0\}$. Let $Y \in L(V : -\infty)$. Then, let $U : L(V) \to L^\alpha(\mathbb{T}, \sigma)$ be the isomorphism that sends V_n to $e^{-in\lambda}$. Therefore, Y can be represented by some f in $L^\alpha(\mathbb{T}, \sigma)$; that is, $Y = \int_{[-\pi,\pi)} f(e^{i\lambda}) \, dZ_V(e^{i\lambda})$. Since $Y \in L(V : -\infty)$, by definition, $Y \in L(V : n)$ for all $n \in \mathbb{Z}$. It follows that $f \in \overline{\text{span}}\{e^{-ik\lambda} : k \leq n\}$ for all $n \in \mathbb{Z}$. Since Fourier coefficients are unique, it follows that $f = 0$. Therefore, $Y = 0$. So, $L(V : -\infty) = \{0\}$ and hence $L(X : -\infty) = \{0\}$. So, by definition, X_n, $n \in \mathbb{Z}$ is regular, as desired. $\qquad\square$

In sharp contrast with the Gaussian case, $\alpha = 2$, where the V_n's in Theorem 4.3.1 are independent, for the non-Gaussian case, $0 < \alpha < 2$, the V_n's are not independent random variables. We state this observation as a proposition.

Proposition 4.3.1. *No (nontrivial) harmonizable non-Gaussian SαS sequences X_n, $n \in \mathbb{Z}$ with $0 < \alpha < 2$ is the moving average of an independent SαS sequence V_n, $n \in \mathbb{Z}$ with $L(X) = L(V)$.*

Proof. Suppose on the contrary that the V_n's are independent. Since $V_n \in L(X)$, $n \in \mathbb{Z}$, V_n is represented by some $f_n \in L^\alpha(\mathbb{T}, \mu)$, $n \in \mathbb{Z}$. Now, as observed above, the mutual independence of the V_n's implies that the f_n's have mutually disjoint supports. We will use E_n to denote the support of f_n, $n \in \mathbb{Z}$. By the correspondence between $L(X)$ and $L^\alpha(\mathbb{T}, \mu)$ and the moving average representation, it follows that

$$e^{-in\lambda} = \sum_{k=0}^{\infty} a_k f_{n-k}(e^{i\lambda}) \text{ in } L^\alpha(\mathbb{T}, \mu).$$

Since X_n, $n \in \mathbb{Z}$ is nontrivial, some f_m, say f_{m_0}, is not identically zero. Then, for all $j \geq 0$,

$$e^{-i(j+m_0)\lambda} = a_j f_{m_0}(e^{i\lambda}) \qquad [\mu]\text{-a.e. on } E_{m_0}.$$

It follows that $a_j \neq 0$ for all $j \geq 0$. Now, putting $j = 0$ and $j = 1$, we get $a_1 = e^{-i\lambda} a_0$ $[\mu]$-a.e. on E_{m_0} and since $\mu(E_{m_0}) > 0$, we get a contradiction. $\qquad\square$

Let $\mu = \mu_a + \mu_s$ be the Lebesgue decomposition of μ relative to σ. Here, μ_a is absolutely continuous with respect to σ and μ_s is singular with respect to σ. We will use $\dfrac{d\mu_a}{d\sigma}$ to denote the Radon-Nikodym derivative of μ_a with respect to σ.

Theorem 4.3.2. *Let X_n, $n \in \mathbb{Z}$ be a harmonizable SαS sequence with $1 < \alpha \leq 2$ and spectral measure μ. Then, X_n, $n \in \mathbb{Z}$ is singular if and only if*

$$\int_{[-\pi,\pi)} \log\left(\frac{d\mu_a}{d\sigma}(e^{i\lambda})\right) d\sigma(e^{i\lambda}) = -\infty.$$

Proof. ($\Rightarrow$) Suppose X_n, $n \in \mathbb{Z}$ is singular. By definition, $L(X : -\infty) =$

$L(X)$. Using the isomorphism that takes X_n to $e^{in\lambda}$ in $L^\alpha(\mathbb{T}, \mu)$, defining $M_n^{(\alpha)} = \overline{span}\{e^{ik\lambda} : k \leq n\}$ and $M_{-\infty}^{(\alpha)} = \cap_n M_n^{(\alpha)}$, it follows that if X_n is singular, then $M_{-\infty}^{(\alpha)} = L^\alpha(\mathbb{T}, \mu)$. Since $1 < \alpha \leq 2$, it follows that $M_n^{(2)} = M_n^{(\alpha)} \cap L^2(\mathbb{T}, \mu) = L^\alpha(\mathbb{T}, \mu) \cap L^2(\mathbb{T}, \mu) = L^2(\mathbb{T}, \mu)$ for all n. It then follows that μ is the measure of some singular weakly stationary sequence, as such from Corollary 1.15.1 it follows that $\int_{[-\pi,\pi)} \log\left(\frac{d\mu_a}{d\sigma}(e^{i\lambda})\right) d\sigma(e^{i\lambda}) = -\infty.$

($\Leftarrow$) Suppose that $\int_{[-\pi,\pi)} \log\left(\frac{d\mu_a}{d\sigma}(e^{i\lambda})\right) d\sigma(e^{i\lambda}) = -\infty$. By Corollary 1.15.1, it follows that $M_n^{(2)} = L^2(\mathbb{T}, \mu)$ for all n. Now $M_n^{(2)} \subseteq M_n^{(\alpha)}$ for all n and $M_n^{(\alpha)}$ is the closure of $M_n^{(2)}$ in $L^\alpha(\mathbb{T}, \mu)$. Therefore, $M_n^{(\alpha)} = L^\alpha(\mathbb{T}, \mu)$ for all n. It follows that X_n, $n \in \mathbb{Z}$ is singular. $\qquad\square$

We now state and prove a Wold-type decomposition for harmonizable SαS sequences.

Theorem 4.3.3 (Wold Decomposition). *Let X_n, $n \in \mathbb{Z}$ be a non-singular harmonizable SαS sequence with $1 < \alpha \leq 2$. Then, there exists a unique decomposition into harmonizable SαS sequences Y_n, $n \in \mathbb{Z}$, Z_n, $n \in \mathbb{Z}$ and V_n, $n \in \mathbb{Z}$ such that*

$$X_n = Y_n + Z_n = \sum_{k=0}^{\infty} a_k V_{n-k} + Z_n,$$

Y_n, $n \in \mathbb{Z}$ is regular, Z_n, $n \in \mathbb{Z}$ is singular and independent of Y_n, $n \in \mathbb{Z}$ and V_n, $n \in \mathbb{Z}$, $a_0 > 0$, and the V_n's are orthogonal. Futhermore, $L(X : n) = L(V : n) + L(X : -\infty)$.

We observe that $L(X : -\infty) = L(Z)$ is independent of $L(V) = L(Y)$, and Z_n is the best approximate of X_n in $L(X : -\infty)$.

Proof. Let E_a denote the set where μ_a, the absolutely continuous part of μ with respect to σ, is concentrated and let E_s denote the set where μ_s, the singular part of μ with respect to σ, is concentrated. These sets can be chosen so that $E_a \cup E_s = \mathbb{T}$ and $E_a \cap E_s = \emptyset$. Now, define the harmonizable SαS sequences

$$Y_n = \int_{[-\pi,\pi)} e^{-in\lambda} \mathbf{1}_{E_a}(e^{i\lambda}) \, d\mathcal{Z}(e^{i\lambda})$$

and

$$Z_n = \int_{[-\pi,\pi)} e^{-in\lambda} \mathbf{1}_{E_s}(e^{i\lambda}) \, d\mathcal{Z}(e^{i\lambda}).$$

Since $E_a \cap E_s = \emptyset$, it follows that the Y_n's and the Z_n's are independent. Now,

Y_n, $n \in \mathbb{Z}$ has spectral measure μ_a and hence by Theorem 4.3.1, Y_n, $n \in \mathbb{Z}$ is regular with

$$Y_n = \sum_{k=0}^{\infty} a_k V_{n-k},$$

where $a_0 > 0$, the V_n's are orthogonal and are independent of the Z_n's. Since the Z_n's have spectral measure μ_s, it follows from Theorem 4.3.2 that the Z_n's are singular. Finally, we observe that $L(X : n) = L(Y : n) + L(Z : n) = L(V : n) + L(Z) = L(V : n) + L(X : -\infty)$. $\qquad\square$

4.4   Invariant Subspaces of $L^p(\mathbb{T}^2)$

Let $\mathbb{T}^2 = \mathbb{T} \times \mathbb{T}$. We will write $L^p(\mathbb{T}^2)$, for $1 < p < \infty$, to denote the collection of all Lebesgue measurable functions $f : \mathbb{T}^2 \to \mathbb{C}$ such that

$$\int_{[-\pi,\pi)} \int_{[-\pi,\pi)} \left| f(e^{i\lambda}, e^{i\theta}) \right|^p \, d\sigma^2(e^{i\lambda}, e^{i\theta}) < \infty.$$

As always, we will identify functions that are equal $[\sigma^2]$-a.e., where σ^2 denotes normalized Lebesgue measure on $\mathbb{T}^2$. Finally, $L^\infty(\mathbb{T}^2)$ will denote the collection of $L^p(\mathbb{T}^2)$ functions that are essentially bounded. Some important subspaces of $L^p(\mathbb{T}^2)$ that will come into play are

$$H^p(\mathbb{T}^2) = \overline{span} \left\{ e^{im\lambda} e^{in\theta} : m, n \geq 0 \right\},$$

$$H_1^p(\mathbb{T}^2) = \overline{span} \left\{ e^{im\lambda} e^{in\theta} : m \geq 0, n \in \mathbb{Z} \right\},$$

$$H_2^p(\mathbb{T}^2) = \overline{span} \left\{ e^{im\lambda} e^{in\theta} : n \geq 0, m \in \mathbb{Z} \right\},$$

$$L_\theta^p(\mathbb{T}) = \overline{span} \{ e^{in\theta} : n \in \mathbb{Z} \},$$

and

$$L_\lambda^p(\mathbb{T}) = \overline{span} \{ e^{in\lambda} : n \in \mathbb{Z} \}.$$

Let $S_1 : L^p(\mathbb{T}^2) \to L^p(\mathbb{T}^2)$ be the linear operator defined by

$$S_1(f)(e^{i\lambda}, e^{i\theta}) = e^{i\lambda} f(e^{i\lambda}, e^{i\theta})$$

and let $S_2 : L^p(\mathbb{T}^2) \to L^p(\mathbb{T}^2)$ be the linear operator defined by

$$S_2(f)(e^{i\lambda}, e^{i\theta}) = e^{i\theta} f(e^{i\lambda}, e^{i\theta}).$$

It is straightforward to verify that S_1 and S_2 are isometries in $\mathcal{L}(L^p(\mathbb{T}^2))$. A subspace $\mathcal{M}$ of $L^p(\mathbb{T}^2)$ is said to be S-invariant if $\mathcal{M}$ is S_1-invariant and S_2-invariant.

A majority of the results from this section are found in [33] and we follow the approach given in that paper. Before we state the main results of this section, we present some preliminary results needed for their proofs. Let $RP(\mathbb{U}^2)$ denote the class of all functions in $\mathbb{U}^2$ which are the real parts of holomorphic functions. Here, $\mathbb{U}$ denotes the unit disk in the complex plane. Hence, $\mathbb{U}^2$ denotes the unit bi-disk.

Theorem 4.4.1 (Rudin [35]). *Suppose f is a lower semicontinuous (l.s.c.) positive function on $\mathbb{T}^2$ and $f \in L^1(\mathbb{T}^2)$. Then there exists a singular (complex Borel) measure μ on $\mathbb{T}^2$, $\mu \geq 0$, such that $P[f - d\mu] \in RP(\mathbb{U}^2)$.*

In the above theorem, $P[f - d\mu]$ denotes the Poisson Integral of $f - d\mu$. We also recall that a function $f : \mathbb{T}^2 \to \mathbb{R}$ is called lower semicontinuous if $\{(e^{i\lambda}, e^{i\theta}) : f(e^{i\lambda}, e^{i\theta}) > \alpha\}$ is open for every real number α. To prove our main results, we need a variation of this theorem, which proves to be a corollary.

Corollary 4.4.1. *Suppose f is real-valued on $\mathbb{T}^2$ and $f \in L^1(\mathbb{T}^2)$. Then there exists a singular (complex Borel) measure μ on $\mathbb{T}^2$, such that $P[f - d\mu] \in RP(\mathbb{U}^2)$.*

We use the following lemma to prove this corollary.

Lemma 4.4.1. *Suppose f is real-valued on $\mathbb{T}^2$ and $f \in L^p(\mathbb{T}^2)$ for $1 \leq p < \infty$. Then there exists two positive l.s.c. functions g_1 and g_2 in $L^p(\mathbb{T}^2)$ such that $f = g_1 - g_2$ $[\sigma^2]$-a.e. on $\mathbb{T}^2$.*

We only need this lemma for the case $p = 1$, but it is no more difficult to prove it for $1 \leq p < \infty$.

Proof. Since f is real-valued on $\mathbb{T}^2$, $f \in L^p(\mathbb{T}^2)$ and continuous functions are dense in $L^p(\mathbb{T}^2)$ there exists ϕ_1 continuous such that

$$\|f - \phi_1\|_p < 2^{-1}$$

and by the reverse triangle inequality we get

$$\|\phi_1\|_p < \left(1 + 2\|f\|_p\right) \cdot 2^{-1}.$$

Now we can find ϕ_2 continuous such that

$$\|(f - \phi_1) - \phi_2\|_p < 2^{-2}$$

and by the reverse triangle inequality we get

$$\|\phi_2\|_p < 2^{-2} + \|f - \phi_1\|_p < 3 \cdot 2^{-2}.$$

Continuing in the manner we get the existence of a sequence of real-valued continuous functions $(\phi_n)_n$ such that

$$f = \sum_{n=1}^{\infty} \phi_n$$

in $L^p(\mathbb{T}^2)$ and

$$\|\phi_n\|_p < C \cdot 2^{-n} \quad \text{for all } n, \text{ where } C = \max\{1 + 2\|f\|_p, \ 3\}.$$

Now, for $\epsilon > 0$, define

$$\psi_n^+ = (\phi_n \vee 0) + \epsilon \cdot 2^{-n}$$

and

$$\psi_n^- = (-\phi_n \vee 0) + \epsilon \cdot 2^{-n}.$$

Then ψ_n^+ and ψ_n^- are positive continuous functions with $\phi_n = \psi_n^+ - \psi_n^-$. So

$$f = \sum_{n=1}^{\infty}(\psi_n^+ - \psi_n^-) = \sum_{n=1}^{\infty}\psi_n^+ - \sum_{n=1}^{\infty}\psi_n^- \quad \text{in } L^p(\mathbb{T}^2).$$

Since

$$\sum_{n=1}^{\infty}\|\psi_n^+\|_p \ \leq \ \sum_{n=1}^{\infty}(\|\phi_n \vee 0\|_p + \epsilon \cdot 2^{-n}) \ \leq \ \sum_{n=1}^{\infty}(\|\phi_n\|_p + \epsilon \cdot 2^{-n})$$

$$< \ \sum_{n=1}^{\infty}(C \cdot 2^{-n} + \epsilon \cdot 2^{-n}) \ < \ \infty$$

we get that there exists a g_1 in $L^p(\mathbb{T}^2)$ such that

$$g_1 = \sum_{n=1}^{\infty}\psi_n^+ \quad \text{in } L^p(\mathbb{T}^2).$$

Similarly, we get that there exists a g_2 in $L^p(\mathbb{T}^2)$ such that

$$g_2 = \sum_{n=1}^{\infty}\psi_n^- \quad \text{in } L^p(\mathbb{T}^2).$$

So we have that

$$f = g_1 - g_2 \quad \text{in } L^p(\mathbb{T}^2).$$

It is left to show that g_1 and g_2 are equal to positive *l.s.c.* functions $[\sigma^2]$-a.e. Let

$$s_n = \sum_{k=1}^{n}\psi_k^+.$$

Since s_n converges to g_1 in $L^p(\mathbb{T}^2)$, there exists a subsequence that converges to g_1 $[\sigma^2]$-a.e. But since s_n is monotone increasing, we get that s_n converges to g_1 $[\sigma^2]$-a.e. and further that $\sup s_n = \lim s_n$. We conclude that $\sup s_n$ is *l.s.c* since the sup of a sequence of continuous functions is *l.s.c.* It is clear that $\sup s_n$ is positive. Therefore, g_1 is equal to a positive *l.s.c.* function $[\sigma^2]$-a.e. Similarly, we get that g_2 is equal to a positive *l.s.c.* function $[\sigma^2]$-a.e. So f is equal $[\sigma^2]$-a.e. to the difference of two positive *l.s.c.* functions. $\square$

We now prove corollary 4.4.1.

Proof. If f is real-valued on $\mathbb{T}^2$ and $f \in L^1(\mathbb{T}^2)$, then Lemma 4.4.1 asserts the existence of two positive *l.s.c.* functions g_1 and g_2 in $L^1(\mathbb{T}^2)$ such that $f = g_1 - g_2$ $[\sigma^2]$-a.e. By Theorem 4.4.1 there exists nonnegative singular measures μ_1 and μ_2 such that $P[g_1 - d\mu_1]$ and $P[g_2 - d\mu_2]$ are in $RP(\mathbb{U}^2)$. Letting $\mu = \mu_1 - \mu_2$ we get a singular measure such that

$$
\begin{aligned}
P[f - d\mu] &= P[(g_1 - g_2) - d(\mu_1 - \mu_2)] \\
&= P[(g_1 - d\mu_1) - (g_2 - d\mu_2)] \\
&= P[g_1 - d\mu_1] - P[g_2 - d\mu_2].
\end{aligned}
$$

So, $P[f - d\mu]$ is in $RP(\mathbb{U}^2)$. $\qquad\square$

We now state and prove the main results.

Theorem 4.4.2. *Let $\mathcal{M} \neq \{0\}$ be a subspace of $L^p(\mathbb{T}^2)$, $1 \leq p < 2$, invariant under S_1 and S_2. Then $\mathcal{M} = qH^p(\mathbb{T}^2)$ where q is a unimodular function if and only if S_1 and S_2 are doubly commuting shifts on $\mathcal{M} \cap L^2(\mathbb{T}^2)$.*

Proof. Let N denote $\mathcal{M} \cap L^2(\mathbb{T}^2)$. Then N is a (closed) invariant subspace of $L^2(\mathbb{T}^2)$ and by hypothesis S_1 and S_2 are doubly commuting shifts on N. Therefore, by Theorem 3.6.5, $N = qH^2(\mathbb{T}^2)$ where q is a unimodular function. Now since N is contained in $\mathcal{M}$ and $\mathcal{M}$ is closed, the closure of N in $L^p(\mathbb{T}^2)$, which is $qH^p(\mathbb{T}^2)$, is contained in $\mathcal{M}$. So we need to show that N is dense in $\mathcal{M}$. To do this, let $f \in \mathcal{M}$, f not identically zero. Then define

$$
u_n = \begin{cases} 0, & |f| \leq n, \\ \log|f|^{-1}, & |f| > n. \end{cases}
$$

Note that $u_n \in L^p(\mathbb{T}^2)$ for all n since

$$
\begin{aligned}
\int |u_n|^p \, dm &= \int_{|f|>n} |\log|f|^{-1}|^p \, dm = \int_{|f|>n} |\log|f||^p \, dm \\
&\leq \int_{|f|>n} |f|^p \, dm \leq \|f\|_p^p < \infty.
\end{aligned}
$$

So in particular, $u_n \in L^1(\mathbb{T}^2)$ and real valued for all n. So by Corollary 4.4.1, there exists a sequence $\{\mu_n\}_{n \geq 0}$ of singular measures such that $P[u_n - d\mu_n] \in RP(\mathbb{U}^2)$ for all n. So there exists a sequence of analytic functions $(F_n)_n$ such that $Re(F_n) = P[u_n - d\mu_n]$. By the M. Riesz theorem, which holds on the polydisc (see [34]), we have $\|F_n\|_p \leq C_p \|u_n\|_p$ for all n. Now since $u_n \in L^p(\mathbb{T}^2)$ and u_n converges to 0 in $L^p(\mathbb{T}^2)$, we get F_n converges to 0 in $L^p(\mathbb{T}^2)$ and hence at least a subsequence converges to zero $[\sigma^2]$-a.e. Let $\phi_n = \exp\{F_n\}$. Then

$$
|\phi_n| = \begin{cases} 1, & |f| \leq n, \\ |f|^{-1}, & |f| > n \end{cases}
$$

and ϕ_n tends to the constant function 1. By construction, $\phi_n f$ is a bounded function dominated by f for all n. Also, $\phi_n f \in \mathcal{M}$ because ϕ_n is bounded analytic and hence is boundedly the limit of analytic trigonometric polynomials. Since $\phi_n f$ is bounded, it is in N. As n goes to infinity $\phi_n f$ converges to f in $L^p(\mathbb{T}^2)$ by the dominated convergence theorem. So each f in $\mathcal{M}$ is the limit of functions from N. So N is dense in $\mathcal{M}$ as desired.

Conversely, if $\mathcal{M} = qH^p(\mathbb{T}^2)$ with q unimodular, then $\mathcal{M} \cap L^2(\mathbb{T}^2) = qH^2(\mathbb{T}^2)$. So S_1 and S_2 are doubly commuting shifts on $\mathcal{M} \cap L^2(\mathbb{T}^2)$ by Theorem 3.6.5. $\qquad\square$

We use the notation $H_o^p(\mathbb{T}^2) = \left\{ f \in H^p(\mathbb{T}^2) : \hat{f}(0,0) = 0 \right\}$ in the next theorem.

Theorem 4.4.3. *Let $\mathcal{M} \neq \{0\}$ be a subspace[1] of $L^p(\mathbb{T}^2)$, $2 < p \leq \infty$, invariant under S_1 and S_2. Then $\mathcal{M} = qH_o^p(\mathbb{T}^2)$ where q is a unimodular function if and only if S_1 and S_2 are doubly commuting shifts on $A(\mathcal{M}) \cap L^2(\mathbb{T}^2)$.*

Here, $A(\mathcal{M})$ means the annihilator of $\mathcal{M}$. That is,

$$A(\mathcal{M}) = \left\{ f \in L^{\frac{p}{p-1}}(\mathbb{T}^2) : \int_{\mathbb{T}^2} fg\, d\sigma^2 = 0, \forall g \in \mathcal{M} \right\}.$$

Proof. If $\mathcal{M} = qH_o^p(\mathbb{T}^2)$ where q is a unimodular function, then $A(\mathcal{M}) = \bar{q}H^{\frac{p}{p-1}}(\mathbb{T}^2)$. Therefore, $A(\mathcal{M}) \cap L^2(\mathbb{T}^2) = \bar{q}H^2(\mathbb{T}^2)$. It then follows from Theorem 3.6.5 that S_1 and S_2 are doubly commuting shifts on $A(\mathcal{M}) \cap L^2(\mathbb{T}^2)$. Conversely, if S_1 and S_2 are doubly commuting shifts on $A(\mathcal{M}) \cap L^2(\mathbb{T}^2)$, then by Theorem 4.4.2 we get that $A(\mathcal{M}) = qH^{\frac{p}{p-1}}(\mathbb{T}^2)$ where q is a unimodular function. Therefore, $\mathcal{M} = \bar{q}H_o^p(\mathbb{T}^2)$ where q is a unimodular function. When $p = \infty$ we need that $\mathcal{M}$ is star-closed to make our final conclusion. $\qquad\square$

We then get the following two corollaries which follow directly from these two theorems.

Corollary 4.4.2. *Let $\mathcal{M} \neq \{0\}$ be a subspace of $H^p(\mathbb{T}^2)$, $1 \leq p < 2$, invariant under S_1 and S_2. Then $\mathcal{M} = qH^p(\mathbb{T}^2)$ where q is an inner function if and only if S_1 and S_2 are doubly commuting on $\mathcal{M} \cap H^2(\mathbb{T}^2)$.*

Corollary 4.4.3. *Let $\mathcal{M} \neq \{0\}$ be a subspace[2] of $H^p(\mathbb{T}^2)$, $2 < p \leq \infty$, invariant under S_1 and S_2. Then $\mathcal{M} = qH_o^p(\mathbb{T}^2)$ where q is an inner function if and only if S_1 and S_2 are doubly commuting shifts on $A(\mathcal{M}) \cap L^2(\mathbb{T}^2)$.*

We shall call a subspace $\mathcal{M}$ of $L^p(\mathbb{T}^2)$ doubly invariant if $\mathcal{M}$ is S-invariant, $S_1(\mathcal{M}) = \mathcal{M}$ and $S_2(\mathcal{M}) = \mathcal{M}$.

Theorem 4.4.4. *Every doubly invariant subspace of $L^p(\mathbb{T}^2)$ is of the form $1_E L^p(\mathbb{T}^2)$, where E is a measurable subset of $\mathbb{T}^2$ and 1_E is the indicator function of E.*

---

[1] Assume further star-closed when $p = \infty$.
[2] Assume further star-closed when $p = \infty$.

It is straightforward to verify that $\mathbf{1}_E L^p(\mathbb{T}^2)$, where E is a measurable subset of $\mathbb{T}^2$ is a doubly invariant subspace of $L^p(\mathbb{T}^2)$. For the other direction, let us call our doubly invariant subspace $\mathcal{M}$. Suppose first that $1 \leq p < 2$. Then, $\mathcal{M} \cap L^2(\mathbb{T}^2)$ is a doubly invariant subspace of $L^2(\mathbb{T}^2)$. Therefore, we know from Theorem 3.6.1, $\mathcal{M} \cap L^2(\mathbb{T}^2) = \mathbf{1}_E L^2(\mathbb{T}^2)$, where E is a measurable subset of $\mathbb{T}^2$. The closure of $\mathbf{1}_E L^2(\mathbb{T}^2)$ in $L^p(\mathbb{T}^2)$ is $\mathbf{1}_E L^p(\mathbb{T}^2)$. So, it remains to show that the closure of $\mathcal{M} \cap L^2(\mathbb{T}^2)$ in $L^p(\mathbb{T}^2)$ is $\mathcal{M}$. This can be accomplished in the same way as we did above. So, $\mathcal{M} = \mathbf{1}_E L^p(\mathbb{T}^2)$, where E is a measurable subset of $\mathbb{T}^2$, as desired. It remains to consider the case when $2 < p \leq \infty$. For this case, as above, use the annihilator. It is straightforward to verify that $A(\mathcal{M})$ is a doubly invariant subspace of $L^{p/(p-1)}(\mathbb{T}^2)$ and that $1 \leq p/(p-1) < 2$. So, we have from above that $A(\mathcal{M}) = \mathbf{1}_E L^{p/(p-1)}(\mathbb{T}^2)$, where E is a measurable subset of $\mathbb{T}^2$. From this, it follows that $\mathcal{M} = \mathbf{1}_{E^c} L^p(\mathbb{T}^2)$ and our proof is complete. The ideas from this argument can be employed, in much the same way, to prove the following two theorems.

Theorem 4.4.5. *Every S-invariant subspace $\mathcal{M}$ of $L^p(\mathbb{T}^2)$ for which $S_1(\mathcal{M}) \subsetneq \mathcal{M}$ and $S_2(\mathcal{M}) = \mathcal{M}$ is of the form*

$$\left(\sum_{j=1}^{\infty} \oplus u_j(e^{i\lambda}, e^{i\theta}) \mathbf{1}_{\mathbb{T} \times K_j} H_1^p(\mathbb{T}^2) \right) \bigoplus \mathbf{1}_E L^p(\mathbb{T}^2),$$

where the u_j's are unimodular, the K_j's are measurable subsets of $\mathbb{T}$ with the property that $\sigma(K_j \cap K_l) = 0$ for all $j \neq l$, where σ denotes normalized Lebesgue measure on $\mathbb{T}$, and E is a measurable subset of $\mathbb{T}^2$ with the property that $\sigma^2 \left(E \cap \left(\mathbb{T} \times \left(\cup_{j=1}^{\infty} K_j \right) \right) \right) = 0$.

Theorem 4.4.6. *Every S-invariant subspace $\mathcal{M}$ of $L^p(\mathbb{T}^2)$ for which $S_1(\mathcal{M}) = \mathcal{M}$ and $S_2(\mathcal{M}) \subsetneq \mathcal{M}$ is of the form*

$$\left(\sum_{j=1}^{\infty} \oplus u_j(e^{i\lambda}, e^{i\theta}) \mathbf{1}_{K_j \times \mathbb{T}} H_2^p(\mathbb{T}^2) \right) \bigoplus \mathbf{1}_E L^p(\mathbb{T}^2),$$

where the u_j's are unimodular, the K_j's are measurable subsets of $\mathbb{T}$ with the property that $\sigma(K_j \cap K_l) = 0$ for all $j \neq l$ and E is a measurable subset of $\mathbb{T}^2$ with the property that $\sigma^2 \left(E \cap \left(\left(\cup_{j=1}^{\infty} K_j \right) \times \mathbb{T} \right) \right) = 0$.

The $\oplus$, in both theorems, denotes the direct sum.

4.5 Harmonizable SαS Fields

In recent years, based on data from finance, insurance and hydrology, it is found that one needs to study the stationary random fields that are not second order $\left(\text{i.e., } E|X_{m,n}|^2 = \infty \right)$. For this, one needs to study the models given by so-called stable fields.

Let $\mathcal{Z}$ is an independently scattered complex isotropic α-stable variable valued set function defined on $\mathcal{B}(\mathbb{T}^2)$, the Borel subsets of $\mathbb{T}^2$. That is, for all disjoint sets $\Delta_1, \cdots, \Delta_n \in \mathcal{B}(\mathbb{T}^2)$, $\mathcal{Z}(\Delta_1), \cdots, \mathcal{Z}(\Delta_n)$ are independent with

$$\varphi_{\mathcal{Z}(\Delta_k)}(t) = \left\{ \begin{array}{ll} \exp\left\{-|t|^\alpha \|\mathcal{Z}(\Delta_k)\|^\alpha\right\} & \text{for } 1 \leq \alpha \leq 2 \\ \exp\left\{-|t|^\alpha \|\mathcal{Z}(\Delta_k)\|\right\} & \text{for } 0 < \alpha < 1 \end{array} \right. , \quad t \in \mathbb{C}.$$

Using $\mathcal{Z}$, we define

$$\mu(\Delta) = \left\{ \begin{array}{ll} \|\mathcal{Z}(\Delta)\|^\alpha & \text{for } 1 \leq \alpha \leq 2 \\ \|\mathcal{Z}(\Delta)\| & \text{for } 0 < \alpha < 1 \end{array} \right. , \quad \text{for } \Delta \in \mathcal{B}(\mathbb{T}^2),$$

and observe that μ is a finite measure defined on $\mathcal{B}(\mathbb{T}^2)$. From this, it follows that if $f \in L^\alpha(\mathbb{T}^2, \mu)$ and $\mathcal{X} = \int_{[-\pi,\pi)} \int_{[-\pi,\pi)} f(e^{i\lambda}, e^{i\theta}) \, d\mathcal{Z}(e^{i\lambda}, e^{i\theta})$, then

$$\varphi_{\mathcal{X}}(t) = \exp\left\{ -|t|^\alpha \int_{[-\pi,\pi)} \int_{[-\pi,\pi)} |f(e^{i\lambda}, e^{i\theta})|^\alpha \, d\mu(e^{i\lambda}, e^{i\theta}) \right\}, \quad t \in \mathbb{C}.$$

A complex random field $X_{m,n}$, $(m,n) \in \mathbb{Z}^2$ is called harmonizable $S\alpha S$ with spectral measure μ, if μ is a finite (positive) measure defined on $\mathcal{B}(\mathbb{T}^2)$ with

$$E \exp\left\{ i \operatorname{Re}\left(\bar{t} \sum_{j=1}^N z_j X_{t_j, l_j} \right) \right\} =$$

$$\exp\left\{ -|t|^\alpha \int_{[\pi,\pi)} \int_{[\pi,\pi)} \left| \sum_{j=1}^N z_j e^{-it_j \lambda - il_j \theta} \right|^\alpha d\mu(e^{i\lambda}, e^{i\theta}) \right\},$$

where $t, z_j \in \mathbb{C}$, and $t_j, l_j \in \mathbb{Z}$, for $j = 1, \cdots, N$. We see from this that $X_{m,n}$, $(m,n) \in \mathbb{Z}^2$ is (strictly) stationary. We may define a harmonizable $S\alpha S$ field equivalently through its spectral representation

$$X_{m,n} = \int_{[-\pi,\pi)} \int_{[-\pi,\pi)} e^{-im\lambda - in\theta} \, d\mathcal{Z}(e^{i\lambda}, e^{i\theta}),$$

where $\mathcal{Z}$ is an independently scattered complex isotropic α-stable variable valued set function defined on $\mathcal{B}(\mathbb{T}^2)$. If $L(X)$ is the closure in probability of the linear span of $X_{m,n}$, $(m,n) \in \mathbb{Z}^2$, then the correspondence between f and

$$\int_{[-\pi,\pi)} \int_{[-\pi,\pi)} f(e^{i\lambda}, e^{i\theta}) \, d\mathcal{Z}(e^{i\lambda}, e^{i\theta})$$

gives an isomorphism between $L^\alpha(\mathbb{T}^2, \mu)$ and $L(X)$, that sends $e^{-im\lambda - in\theta}$ to $X_{m,n}$. Hence, every $Y \in L(X)$ has a representation of the form

$$\int_{[-\pi,\pi)} \int_{[-\pi,\pi)} f(e^{i\lambda}, e^{i\theta}) \, d\mathcal{Z}(e^{i\lambda}, e^{i\theta})$$

for some $f \in L^{\alpha}(\mathbb{T}^2, \mu)$ and has a radially symmetric distribution.

For a harmonizable SαS field $X_{m,n}$, $(m, n) \in \mathbb{Z}^2$, we define the concept of regularity. Let $L(X : m, n)$ denote the closure in probability of the linear span of $\{X_{k,l} : k \leq m, l \leq n\}$ and let $L(X : -\infty) = \cap_{m,n} L(X : m, n)$. $X_{m,n}$, $(m, n) \in \mathbb{Z}^2$ is called **strongly regular** if $L(X : -\infty) = \{0\}$. These definitions are consistent with those given for weakly stationary fields. It will be advantageous to redefine these concepts in $L^{\alpha}(\mathbb{T}^2, \mu)$. Using the isomorphism that takes $X_{m,n}$ to $e^{im\lambda+in\theta}$. Under this isomorphism, it is straightforward to see that $X_{m,n}$, $(m, n) \in \mathbb{Z}^2$ is strongly regular if $M_{-\infty}^{(\alpha)} = \{0\}$, where $M_{-\infty}^{(\alpha)} = \cap_{m,n} M_{m,n}^{(\alpha)}$ and $M_{m,n}^{(\alpha)}$ is equal to the span closure in $L^{\alpha}(\mathbb{T}^2, \mu)$ of $\{e^{ik\lambda+il\theta} : k \leq m, l \leq n\}$. In the following theorem, we assume that, for each (m, n) in $\mathbb{Z}^2$, the σ-algebra generated by $\{X_{k,l} : k \leq m, l \in \mathbb{Z}\}$ is conditionally independent of the σ-algebra generated by $\{X_{k,l} : k \in \mathbb{Z}, l \leq n\}$ given the σ-algebra generated by $\{X_{k,l} : k \leq m, l \leq n\}$.

Theorem 4.5.1. *Let $X_{m,n}$, $(m, n) \in \mathbb{Z}^2$ be a harmonizable SαS field with $1 < \alpha \leq 2$ and spectral measure μ. Then, the following are equivalent.*

    1. $X_{m,n}$, $(m, n) \in \mathbb{Z}^2$ is strongly regular.

    2. Each nonzero function in $M_{0,0}^{(\alpha)}$ is different from zero $[\sigma^2]$-a.e.

Proof. ($\Leftarrow$) Let us suppose that $M_{-\infty}^{(\alpha)} \neq \{0\}$. Let $f \in M_{-\infty}^{(\alpha)}$ with $f \neq 0$. Define

$$\mathcal{M}_f = \overline{span}^{L^{\alpha}(\mathbb{T}, d\mu)} \left\{ e^{im\lambda+in\theta} f(e^{i\lambda}, e^{i\theta}) : (m, n) \in \mathbb{Z}^2 \right\}$$

and

$$\mathcal{N} = \overline{span}^{L^{\alpha}(\mathbb{T}, |f|^{\alpha} d\mu)} \left\{ e^{im\lambda+in\theta} : (m, n) \in \mathbb{Z}^2 \right\}.$$

Finally, defining $T_f : \mathcal{N} \to \mathcal{M}_f$ by $\phi \mapsto \phi f$. T_f is an onto isometry. Since continuous functions are dense in $L^{\alpha}(\mathbb{T}, |f|^{\alpha} d\mu)$, we get that $\mathcal{N} = L^{\alpha}(\mathbb{T}, |f|^{\alpha} d\mu)$. Therefore, $1_B f \in \mathcal{M}_f \subset M_{-\infty}^{(\alpha)} \subset M_{0,0}^{(\alpha)}$ for all $B \in \mathcal{B}(\mathbb{T}^2)$. Thus contradicting our hypothesis.

($\Rightarrow$) Suppose that $X_{m,n}$, $(m, n) \in \mathbb{Z}^2$ is strongly regular; that is, $M_{-\infty}^{(\alpha)} = \{0\}$. Since $M_{-\infty}^{(2)} \subseteq M_{-\infty}^{(\alpha)}$, it follows that $M_{-\infty}^{(2)} = \{0\}$. It is easy to see that the conditional independence assumption implies the strong commuting property on $L^2(\mathbb{T}^2, \mu)$. It then follows from Theorem 2.9.8 that μ is absolutely continuous with respect to σ^2 and $\frac{d\mu}{d\sigma^2} = |\varphi|^2$, where φ is a strongly outer function in $H^2(\mathbb{T}^2)$. Now, using Corollary 2.12.3, we can rewrite $\frac{d\mu}{d\sigma^2} = |\psi|^{\alpha}$, where ψ is a function in $H^{\alpha}(\mathbb{T}^2)$. Then, the mapping $e^{-(im\lambda+in\theta)} \to e^{im\lambda+in\theta} \psi(e^{i\lambda}, e^{i\theta})$, $(m, n \geq 0)$ extends to an isometry from $M_{0,0}^{(\alpha)}$ to $H^{\alpha}(\mathbb{T}^2)$. Since each nonzero function in $H^{\alpha}(\mathbb{T}^2)$ is different from zero $[\sigma^2]$-a.e. and in particular $\psi \neq 0$ $[\sigma^2]$-a.e., every function in $M_{0,0}^{(\alpha)}$ has the same property. So, every nonzero member of $M_{0,0}^{(\alpha)}$ is different from zero $[\sigma^2]$-a.e. $\qquad \square$

4.6   Proper MA Representations and Outer Functions

In the study of proper moving average representations for weakly stationary sequences, it was shown in Section 1.8 that such a representation is equivalent to the spectral measure of the random sequence being absolutely continuous with respect to σ, with the spectral density having the form $|\varphi|^2$, where φ is an outer function in $H^2(\mathbb{T})$. We recall that a function φ in $H^2(\mathbb{T})$ is called outer if

$$\log \left| \int_{[-\pi,\pi)} \varphi(e^{i\lambda})\, d\sigma(e^{i\lambda}) \right| = \int_{[-\pi,\pi)} \log \left| \varphi(e^{i\lambda}) \right|\, d\sigma(e^{i\lambda}).$$

It is well known, see [36], that this condition is equivalent to

$$\overline{\operatorname{span}} \left\{ e^{in\lambda} \varphi(e^{i\lambda}) : n \geq 0 \right\} = H^2(\mathbb{T}). \tag{4.13}$$

We will call any φ in $H^2(\mathbb{T})$ that satisfies Equation (4.13) a **generator** of $H^2(\mathbb{T})$. Using this terminology, one observes that outer functions in $H^2(\mathbb{T})$ can be characterized as generators of $H^2(\mathbb{T})$.

Similarly, in the study of proper moving average representations for weakly stationary random fields with semigroup ordering, it was shown in Section 2.11 that such a representation is equivalent to the spectral measure of the random field being absolutely continuous with respect to σ^2, with the spectral density having the form $|\varphi|^2$, where φ is an outer function in $H_S^2(\mathbb{T}^2, \sigma^2)$. We recall that a function φ in $H_S^2(\mathbb{T}^2, \sigma^2)$ is called **outer** if

$$\log \left| \int_{[-\pi,\pi)} \int_{[-\pi,\pi)} \varphi(e^{i\lambda}, e^{i\theta})\, d\sigma^2(e^{i\lambda}, e^{i\theta}) \right|$$
$$= \int_{[-\pi,\pi)} \int_{[-\pi,\pi)} \log \left| \varphi(e^{i\lambda}, e^{i\theta}) \right|\, d\sigma^2(e^{i\lambda}, e^{i\theta}). \tag{4.14}$$

It is well known, see Theorem 2.10.4, that this condition is equivalent to

$$\overline{\operatorname{span}} \left\{ e^{im\lambda+in\theta} \varphi(e^{i\lambda}, e^{i\theta}) : (m,n) \in S \right\} = H_S^2(\mathbb{T}^2, \sigma^2). \tag{4.15}$$

We will call any φ in $H_S^2(\mathbb{T}^2, \sigma^2)$ that satisfies Equation (4.15) a **generator** of $H_S^2(\mathbb{T}^2, \sigma^2)$. Using this terminology, one observes that outer functions in $H_S^2(\mathbb{T}^2, \sigma^2)$ can be characterized as generators of $H_S^2(\mathbb{T}^2, \sigma^2)$.

Finally, in the study of proper moving average representations for weakly stationary random fields in the quarter-plane, it was shown in Theorem 2.9.8 that such a representation is equivalent to the spectral measure of the random field being absolutely continuous with respect to σ^2, with the spectral density having the form $|\varphi|^2$, where φ is a strongly outer function in $H^2(\mathbb{T}^2)$. We recall that a function φ in $H^2(\mathbb{T}^2)$ is called **strongly outer** if

$$\overline{\operatorname{span}} \left\{ e^{im\lambda+in\theta} \varphi(e^{i\lambda}, e^{i\theta}) : m, n \geq 0 \right\} = H^2(\mathbb{T}^2). \tag{4.16}$$

Here, as before, such a function will be called a **generator** of $H^2(\mathbb{T}^2)$. Unlike the cases above, this condition is **not** equivalent to φ being outer. Here, as in the semigroup case, φ in $H^2(\mathbb{T}^2)$ is called **outer** if it satisfies Equation (4.14).

Rudin showed, in [35], that every strongly outer function is outer. He also showed, in [35], that there are outer functions in $H^2(\mathbb{T}^2)$ that are not strongly outer. These observations confirm the fact that outer functions in $H^2(\mathbb{T}^2)$ can **not** be characterized as generators of $H^2(\mathbb{T}^2)$. In [16], K. Izuchi and Y. Matsugu found that outer functions in $H^2(\mathbb{T}^2)$ can be characterized as generators of some Hardy-type spaces. In what follows, we present some of their work. The following results are also found in [5]. We start by introducing some additional notations and terminology. The notations we introduce are different than that used in [16] and [5], but are more consistent with the notation used in this book.

Let $A \subseteq \mathbb{Z}^2$ and g be in $L^2(\mathbb{T}^2, \sigma^2)$. Then, we define

$$[A] = \overline{\operatorname{span}} \left\{ e^{im\lambda + in\theta} : (m, n) \in A \right\}$$

and

$$[g]_A = \overline{\operatorname{span}} \left\{ e^{im\lambda + in\theta} g(e^{i\lambda}, e^{i\theta}) : (m, n) \in A \right\},$$

where, in both cases, the closure is in $L^2(\mathbb{T}^2, \sigma^2)$.

Now, let $A_\lambda = \{(m, n) : m \geq 0, n \in \mathbb{Z}\}$ and $A_\theta = \{(m, n) : m \in \mathbb{Z}, n \geq 0\}$ and define $H_\lambda^2(\mathbb{T}^2) = [A_\lambda]$ and $H_\theta^2(\mathbb{T}^2) = [A_\theta]$. Also, let $B_\lambda = \{(m, 0) : m \geq 0\}$ and $B_\theta = \{(0, n) : n \geq 0\}$ and define $H^2(\mathbb{T}_\lambda) = [B_\lambda]$ and $H^2(\mathbb{T}_\theta) = [B_\theta]$. Now, for f in $H_\lambda^2(\mathbb{T}^2)$ and θ fixed, we define the **cut function**, $f_\theta(e^{i\lambda}) = f(e^{i\lambda}, e^{i\theta})$. Note that $f_\theta \in H^2(\mathbb{T}_\lambda)$ for $[\sigma]$-a.e. $e^{i\theta}$. We say f in $H_\lambda^2(\mathbb{T}^2)$ is **θ-outer** if f_θ is outer for $[\sigma]$-a.e. $e^{i\theta}$. In an analogous way, we define **λ-outer**.

Theorem 4.6.1. *Let f be in $H_\lambda^2(\mathbb{T}^2)$. Then, f is θ-outer if and only if $[f]_{A_\lambda} = H_\lambda^2(\mathbb{T}^2)$.*

Proof. First, suppose that f is θ-outer, then $f \neq 0$ $[\sigma^2]$-a.e. and by Theorem 3.6.2, $[f]_{A_\lambda} = qH_\lambda^2(\mathbb{T}^2)$, where q is a unimodular function in $H_\lambda^2(\mathbb{T}^2)$. It follows that the cut function q_θ is inner for $[\sigma]$-a.e. $e^{i\theta}$. Since $f = qg$ for some g in $H_\lambda^2(\mathbb{T}^2)$ and the cut function f_θ is outer, it follows that q_θ is constant for $[\sigma]$-a.e. $e^{i\theta}$. If follows that $qH_\lambda^2(\mathbb{T}^2) = H_\lambda^2(\mathbb{T}^2)$. Therefore, $[f]_{A_\lambda} = H_\lambda^2(\mathbb{T}^2)$, as desired.

Now, suppose that $[f]_{A_\lambda} = H_\lambda^2(\mathbb{T}^2)$. As was noted before, $f_\theta \in H^2(\mathbb{T}_\lambda)$ for $[\sigma]$-a.e. $e^{i\theta}$. Let

$$E = \left\{ e^{i\theta} : f_\theta \in H^2(\mathbb{T}_\lambda) \right\}.$$

Now, define, for $e^{i\theta} \in E$,

$$F(z, e^{i\theta}) = \exp \left[\int_{[-\pi, \pi)} \frac{e^{i\lambda} + z}{e^{i\lambda} - z} \log \left(f(e^{i\lambda}, e^{i\theta}) \right) d\sigma(e^{i\lambda}) \right] \qquad |z| < 1,$$

and for $e^{i\theta} \notin E$, define $F(z, e^{i\theta}) = 1$. It is well known, from the theory of function, that for each $e^{i\theta} \in E$, F_θ is an outer function and the radial limits of F_θ exist for $[\sigma]$-a.e. $e^{i\lambda}$ and $|F(e^{i\lambda}, e^{i\theta})| = f(e^{i\lambda}, e^{i\theta})$ for $[\sigma]$-a.e. $e^{i\lambda}$, where $F(e^{i\lambda}, e^{i\theta})$ denotes the radial limit function of $F(z, e^{i\theta})$. It follows that $|F(e^{i\lambda}, e^{i\theta})| = f(e^{i\lambda}, e^{i\theta})$ $[\sigma^2]$-a.e. and if we define $q = \dfrac{f}{F}$, then q is unimodular. We then have that $[f]_{A_\lambda} = q[F]_{A_\lambda}$ and since F is θ-outer, it follows that $[f]_{A_\lambda} = qH_\lambda^2(\mathbb{T}^2)$. Combining this with our supposition gives us that $H_\lambda^2(\mathbb{T}^2) = qH_\lambda^2(\mathbb{T}^2)$. It follows that q_θ is constant for $[\sigma]$-a.e. $e^{i\theta}$. This together with the fact that $f = qF$ and F is θ-outer implies that f is also θ-outer. $\hfill\square$

Now, let $S = \{(m, n) \in \mathbb{Z}^2 : m \geq 1, n \in \mathbb{Z}\} \cup \{(0, n) : n \geq 0\}$. It can be verified that S is a semigroup. If $\mathcal{M}$ is any subspace of $L^2(\mathbb{T}^2, \sigma^2)$ and A is any subset of $\mathbb{Z}^2$, then

$$[\mathcal{M}]_A = \overline{\operatorname{span}}\left\{e^{im\lambda + in\theta}g(e^{i\lambda}, e^{i\theta}) : (m, n) \in A, g \in \mathcal{M}\right\}.$$

If φ is in $H_S^2(\mathbb{T}^2, \sigma^2)$, then, as stated earlier, φ is outer if and only if $[\varphi]_S = [S]$. Using this fact, we make the following observation

$$H_\lambda^2(\mathbb{T}^2) = [A_\lambda] = [[S]]_{A_\lambda} = [[\varphi]_S]_{A_\lambda} = [\varphi]_{A_\lambda}.$$

From this observation and the above theorem, we get the following corollary.

Corollary 4.6.1. *Let f be in $H_S^2(\mathbb{T}^2, \sigma^2)$. If f is outer, then f is θ-outer.*

Since $H^2(\mathbb{T}^2) \subseteq H_S^2(\mathbb{T}^2, \sigma^2)$, the next corollary is immediate.

Corollary 4.6.2. *Let f be in $H^2(\mathbb{T}^2)$. If f is outer, then f is θ-outer.*

We make a few observations before stating our next theorem. Let f be in $H^2(\mathbb{T}^2)$. Then, as noted above, $f_\theta \in H^2(\mathbb{T}_\lambda)$ for $[\sigma]$-a.e. $e^{i\theta}$. Let

$$E = \{\theta : f_\theta \in H^2(\mathbb{T}_\lambda)\}.$$

Since f is in $H^2(\mathbb{T}^2)$, it follows that

$$h(e^{i\theta}) = \int_{[-\pi,\pi)} f(e^{i\lambda}, e^{i\theta})\, d\sigma(e^{i\lambda}) = \sum_{k=0}^{\infty} \hat{f}(0, k)e^{ik\theta}$$

is in $H^2(\mathbb{T}_\theta)$. Note that integration over E is the same as integration over $[-\pi, \pi)$ since $\sigma(E^*) = 1$, where $E^* = \{e^{i\theta} : \theta \in E\}$. We now have the following.

$$\log\left|\int_{[-\pi,\pi)} \int_{[-\pi,\pi)} f(e^{i\lambda}, e^{i\theta})\, d\sigma^2(e^{i\lambda}, e^{i\theta})\right|$$

$$= \log\left|\int_E \int_{[-\pi,\pi)} f(e^{i\lambda}, e^{i\theta})\, d\sigma(e^{i\lambda})\, d\sigma(e^{i\theta})\right|$$

$$\leq \int_E \log \left| \int_{[-\pi,\pi)} f(e^{i\lambda}, e^{i\theta}) \, d\sigma(e^{i\lambda}) \right| \, d\sigma(e^{i\theta})$$

$$\leq \int_E \int_{[-\pi,\pi)} \log \left| f(e^{i\lambda}, e^{i\theta}) \right| \, d\sigma(e^{i\lambda}) \, d\sigma(e^{i\theta})$$

$$= \int_{[-\pi,\pi)} \int_{[-\pi,\pi)} \log \left| f(e^{i\lambda}, e^{i\theta}) \right| \, d\sigma(e^{i\lambda}) \, d\sigma(e^{i\theta}).$$

Note that the first inequality follows from the fact that h is in $H^2(\mathbb{T}_\theta)$ and the second inequality follows from the fact that f_θ is in $H^2(\mathbb{T}_\lambda)$ for all $\theta \in E$. We now make two observations. First, if f is outer, then equality must hold throughout. That implies that both h and f_θ must be outer. On the other hand, if both h and f_θ are outer, then equality would once again hold throughout and f would be outer. These observations show the equivalence of 1. and 3. in the following theorem. The equivalence of 1. and 2. follows from symmetry.

Theorem 4.6.2. *Let f be in $H^2(\mathbb{T}^2)$. Then, the following are equivalent.*

1. f is outer.

2. f is λ-outer and $\displaystyle\sum_{k=0}^{\infty} \hat{f}(k,0)e^{ik\lambda}$ is outer in $H^2(\mathbb{T}_\lambda)$.

3. f is θ-outer and $\displaystyle\sum_{k=0}^{\infty} \hat{f}(0,k)e^{ik\theta}$ is outer in $H^2(\mathbb{T}_\theta)$.

4.7 Remarks and Related Literature

For readers interested in the study of applications of texture identification to engineering problems like pattern recognition and applications to the diagnostic of features of worn out parts, one might refer to [MJ] or [KU].

For those interested in statistical limit theorems for inference in analogue of ARMA random fields one might reference P. Whittle ([41], [42]) and Chapter 8 of the book of M. Rosenblatt referenced in the "Remarks and Related Literature" at the end of Chapter 1.

[MJ] J. Mao and A.K. Jain, *Texture Classification and Segmentation Using Multiresolution Simultaneous Autoregressive Models*, Pattern Recognition, 25, 2, pp. 173–188, 1992.

[KU] Sam-Deuk Kim, S. Udpa, *Texture classification using rotated wavelet filters*, IEEE Transactions on Systems, Man, and Cybernetics - Part A: Systems and Humans (2000), Volume: 30, Issue: 6, 847–852.

A

Background Material

A.1 Projections

The following theorems are well known (see [7]) and are recorded here for reference.

Theorem A.1.1. *A necessary and sufficient condition that the product* $P = P_1 P_2$ *of two projections* P_1 *and* P_2 *be a projection is that* P_1 *commutes with* P_2. *If this condition is satisfied and if the ranges of* P, P_1 *and* P_2 *are* $\mathcal{M}$, $\mathcal{M}_1$ *and* $\mathcal{M}_2$, *respectively, then* $\mathcal{M} = \mathcal{M}_1 \cap \mathcal{M}_2$.

Theorem A.1.2. *A necessary and sufficient condition that the difference* $P = P_1 - P_2$ *of two projections* P_1 *and* P_2 *be a projection is that* $P_2 \leq P_1$. *If this condition is satisfied and if the ranges of* P, P_1 *and* P_2 *are* $\mathcal{M}$, $\mathcal{M}_1$ *and* $\mathcal{M}_2$, *respectively, then* $\mathcal{M} = \mathcal{M}_1 \ominus \mathcal{M}_2$. *Note:* $P_2 \leq P_1$ *is equivalent to* $\mathcal{M}_2 \subseteq \mathcal{M}_1$.

Theorem A.1.3. *If* P *is an operator and if* $\{P_j\}$ *is a family of projections such that* $\sum_j P_j = P$, *then a necessary and sufficient condition that* P *be a projection is that* $P_j \perp P_k$ *whenever* $j \neq k$. *If this condition is satisfied and if, for each* j, *the range of* P_j *is the subspace* $\mathcal{M}_j$, *then the range of* P *is* $\sum_j \oplus \mathcal{M}_j$.

Theorem A.1.4. *Let* $(P_n)_{n=1}^{\infty}$ *be a monotone decreasing sequence of projections defined on a Hilbert space* $\mathcal{H}$. *Then the sequence* $(P_n)_{n=1}^{\infty}$ *converges strongly to an operator* P, *which is a projection defined on* $\mathcal{H}$ *with range* $\bigcap_{n=1}^{\infty} P_n(\mathcal{H})$.

Lemma A.1.1. *Let* $\mathcal{H}$ *be a Hilbert space and let* M *and* N *be subspaces of* $\mathcal{H}$ *such that* P_M *and* P_N *commute. If we write* $M = \overline{span}\{x_i : i \in I\}$ *for some set* I *and let* $z_i = P_N x_i$, *we get* $J = \overline{span}\{z_i : i \in i\} = M \cap N$ *and* $P_J = P_M P_N$.

Proof. By definition, $P_M x_i = x_i$ for all $i \in I$ and by the commutativity of P_M and P_N and the definition of z_i, it follows that $z_i = P_N P_M x_i = P_{M \cap N} x_i$ for all $i \in I$. Therefore, $z_i \in M \cap N$ for all $i \in I$. That is, $J \subseteq M \cap N$. To see that $J = M \cap N$, suppose that there exists a $y \in M \cap N$ such that $(y, z_i)_{\mathcal{H}} = 0$ for all $i \in I$. Since $y \in N$, $P_N y = y$ and so

$$0 = (y, z_i)_{\mathcal{H}} = (y, P_N x_i)_{\mathcal{H}} = (P_N y, x_i)_{\mathcal{H}} = (y, x_i)_{\mathcal{H}}$$

for all $i \in I$. This implies that $y \perp M$ and yet $y \in M$. Therefore, $y = 0$ and so $J = M \cap N$ as desired. $\qquad\qquad\qquad\qquad\qquad\qquad\qquad\qquad\qquad\qquad\qquad\square$

A.2   Orthogonally Scattered Set Functions

In this section, we review the basic fundamentals of orthogonally scattered set functions and stochastic integration. These topics will be used in the development of the spectral representation of a weakly stationary random field.

Let (Λ, Σ) be a measurable space. Let $Z : \Sigma \to \mathcal{H}$ be a set function, where $\mathcal{H}$ is a Hilbert space. Z is called orthogonally scattered, if

1. for any disjoint $\Delta_1, \Delta_2 \in \Sigma$, $(Z(\Delta_1), Z(\Delta_2))_{\mathcal{H}} = 0$ and
2. for disjoint $\{\Delta_n\}_{n=1}^{\infty} \subseteq \Sigma$, $Z(\cup_{n=1}^{\infty} \Delta_n) = \sum_{n=1}^{\infty} Z(\Delta_n)$.

Let Z be orthogonally scattered and let $F(\Delta) = \|Z(\Delta)\|_{\mathcal{H}}^2$. F is a finite positive measure defined on Σ. Note that if $\{\Delta_n\}_{n=1}^{\infty} \subseteq \Sigma$ is a disjoint collection, then

$$
\begin{aligned}
F(\cup_{k=1}^{\infty} \Delta_k) &= \|Z(\cup_{k=1}^{\infty} \Delta_k)\|_{\mathcal{H}}^2 \\
&= \left\| \sum_{k=1}^{\infty} Z(\Delta_k) \right\|_{\mathcal{H}}^2 \\
&= \sum_{k=1}^{\infty} \|Z(\Delta_k)\|_{\mathcal{H}}^2 \\
&= \sum_{k=1}^{\infty} F(\Delta_k).
\end{aligned}
$$

This measure F is called the control measure for Z.

For $f \in L^2(\Lambda, \Sigma, F)$, we will now define the stochastic integral of f with respect to the orthogonally scattered set function Z,

$$
\int_{\Lambda} f(\lambda) \, dZ(\lambda).
$$

Hereafter, we will write $L^2(F)$ instead of $L^2(\Lambda, \Sigma, F)$. We start by defining this integral for simple functions. Let $\Delta_1, \cdots, \Delta_n$ be a disjoint collection of sets in Σ and let $f_1, \cdots, f_n$ be complex numbers. Then, $f = \sum_{k=1}^{n} f_k \chi_{\Delta_k}$ is a simple function. Note χ_{Δ} denotes the characteristic function of Δ. For such a simple function, we define

$$
\int_{\Lambda} f(\lambda) \, dZ(\lambda) = \sum_{k=1}^{n} f_k Z(\Delta_k).
$$

Note then that

$$\left\| \int_\Lambda f(\lambda)\, dZ(\lambda) \right\|_{\mathcal{H}}^2 = \left\| \sum_{k=1}^n f_k Z(\Delta_k) \right\|_{\mathcal{H}}^2$$

$$= \sum_{k=1}^n |f_k|^2\, \|Z(\Delta_k)\|_{\mathcal{H}}^2$$

$$= \sum_{k=1}^n |f_k|^2 F(\Delta_k)$$

$$= \int_\Lambda |f(\lambda)|^2\, dF(\lambda).$$

Now, for any $f \in L^2(F)$, there exists a sequence of simple functions $(f_n)_{n=1}^\infty$ such that f_n converges to f in $L^2(F)$. Therefore, using the same idea as above, we see that

$$\left\| \int_\Lambda f_n(\lambda)\, dZ(\lambda) - \int_\Lambda f_m(\lambda)\, dZ(\lambda) \right\|_{\mathcal{H}}^2 = \int_\Lambda |f_n(\lambda) - f_m(\lambda)|^2\, dF(\lambda).$$

It follows from this equation and the fact that $(f_n)_{n=1}^\infty$ is Cauchy in $L^2(F)$ that $\left(\int_\Lambda f_n(\lambda)\, dZ(\lambda) \right)_{n=1}^\infty$ is Cauchy in $\mathcal{H}$ and consequently has a limit in $\mathcal{H}$. This limit is called the stochastic integral of f with respect to the orthogonally scattered set function Z. We point out that this limit is independent of our choice of simple functions. This integral has the following properties:

1. $\int_\Lambda (a_1 f_1(\lambda) + a_2 f_2(\lambda))\, dZ(\lambda) = a_1 \int_\Lambda f_1(\lambda)\, dZ(\lambda) + a_2 \int_\Lambda f_2(\lambda)\, dZ(\lambda)$
   for all a_1 and a_2 scalars and f_1 and f_2 in $L^2(F)$,

2. If $f_n, f \in L^2(F)$, $n \in \mathbb{N}$, with f_n converging to f in $L^2(F)$, then $\int_\Lambda f_n(\lambda)\, dZ(\lambda)$ converges to $\int_\Lambda f(\lambda)\, dZ(\lambda)$ in $\mathcal{H}$,

3. $\left\| \int_\Lambda f(\lambda)\, dZ(\lambda) \right\|_{\mathcal{H}}^2 = \int_\Lambda |f(\lambda)|^2 dF(\lambda)$ for all $f \in L^2(F)$, and

4. $\left(\int_\Lambda f(\lambda) dZ(\lambda), \int_\Lambda g(\lambda) dZ(\lambda) \right)_{\mathcal{H}} = \int_\Lambda f(\lambda)\overline{g(\lambda)}\, dF(\lambda)$ for all $f, g \in L^2(F)$.

A.3 Representation Theorems

In this section, we present some of the background needed to understand the development of the spectral representation of a weakly stationary random field. This material can be found in the same format in Walter Rudin's book [37]. It is included here for completeness.

Let $\mathcal{M}$ be a σ-algebra of subsets of some set Ω, and let H be a Hilbert space. The resolution of the identity (on $\mathcal{M}$) is the mapping

$$\mathcal{E} : \mathcal{M} \to \mathcal{L}(H),$$

where $\mathcal{L}(H)$ denotes the collection of all bounded linear operators on H, with the following properties:

1. $\mathcal{E}(\emptyset) = 0$ and $\mathcal{E}(\Omega) = I$ (the identity).
2. Each $\mathcal{E}(\Delta)$ is a self-adjoint projection.
3. $\mathcal{E}(\Delta_1 \cap \Delta_2) = \mathcal{E}(\Delta_1)\mathcal{E}(\Delta_2)$.
4. If $\Delta_1 \cap \Delta_2 = \emptyset$, then $\mathcal{E}(\Delta_1 \cup \Delta_2) = \mathcal{E}(\Delta_1) + \mathcal{E}(\Delta_2)$.
5. For every ζ and η in H, the set function $\mathcal{E}_{\zeta,\eta}$ defined by $\mathcal{E}_{\zeta,\eta}(\Delta) = (\mathcal{E}(\Delta)\zeta, \eta)$ is a complex measure on $\mathcal{M}$.

We make two observations that will be used later. First, since $E(\Delta)$ is a self-adjoint projection, we have that

$$\mathcal{E}_{\zeta,\zeta}(\Delta) = (\mathcal{E}(\Delta)\zeta, \zeta) = \|\mathcal{E}(\Delta)\zeta\|^2, \quad (\zeta \in H).$$

Therefore, for every $\zeta \in H$, $\mathcal{E}_{\zeta,\zeta}$ is a finite positive measure on $\mathcal{M}$. Finally, we see from this that $E_{\zeta,\zeta}(\Omega) = \|\zeta\|^2$.

Now, let $\{\Delta_n\}_{n=1}^{\infty}$ be a disjoint sequence of sets in $\mathcal{M}$. Then, for $m \neq n$, it follows from the properties of $\mathcal{E}$ that

$$(\mathcal{E}(\Delta_n)\zeta, \mathcal{E}(\Delta_m)\zeta)_H = (\mathcal{E}(\Delta_m)\mathcal{E}(\Delta_n)\zeta, \zeta)_H = (\mathcal{E}(\Delta_m \cap \Delta_n)\zeta, \zeta)_H = 0.$$

That is, $\mathcal{E}(\Delta_n)\zeta$ and $\mathcal{E}(\Delta_m)\zeta$ are orthogonal when $n \neq m$. Further, it follows from the properties of $\mathcal{E}$ that

$$(\mathcal{E}(\cup_{n=1}^{\infty}\Delta_n)\zeta, \eta)_H = \sum_{n=1}^{\infty}(\mathcal{E}(\Delta_n)\zeta, \eta)_H, \text{ for all } \eta \in H.$$

It then follows from this that

$$\mathcal{E}(\cup_{n=1}^{\infty}\Delta_n)\zeta = \sum_{n=1}^{\infty}\mathcal{E}(\Delta_n)\zeta$$

in the norm topology on H.

These observations show that for $\zeta \in H$, $Z : \mathcal{M} \to H$, defined by

$$Z(\Delta) = \mathcal{E}(\Delta)\zeta, \quad (\Delta \in \mathcal{M})$$

is an orthogonally scattered set function on $\mathcal{M}$, with control measure $\mathcal{E}_{\zeta,\zeta}$.

We finish this section by giving a representation theorem for a pair of commuting unitary operators. These results can be used to develop the spectral representation of a weakly stationary random field.

Let U_1 and U_2 be two commuting unitary operators in $\mathcal{L}(H)$. Let A be the commutative B^*-algebra in $\mathcal{L}(H)$ generated by U_1 and U_2. The polynomials in U_1, U_1^*, U_2, U_2^* are dense in A. If Δ is the maximal ideal space of A, then we can identify Δ with a compact subset of $\mathbb{T}^2$. To see this, define

$$\phi(h) = (\hat{U}_1(h), \hat{U}_2(h)), \qquad (h \in \Delta),$$

where $\hat{x}$ denotes the Gelfand transform of x. Then, ϕ is a homeomorphism of Δ onto a compact set K in $\mathbb{T}^2$. We can transfer $\hat{A}$ from Δ to K and can regard K as the maximal ideal space of A. To make this precise, define

$$\psi(x) = \hat{x} \circ \phi^{-1}, \qquad (x \in A).$$

Since A is semisimple, ψ is an isomorphism from A to $C(K)$. One can now verify that

$$\psi(U_1^m U_2^n)(e^{i\lambda}, e^{i\theta}) = e^{im\lambda + in\theta}, \tag{A.1}$$

for $m, n \in \mathbb{Z}$.

It now follows from Theorem 12.22 of [37] that there exists a unique resolution of the identity E on $\mathcal{B}(K)$ which satisfies

$$T = \int_K \psi(T) \, dE, \qquad (T \in A).$$

Henceforth, we will write

$$T = \int_{[-\pi, \pi)} \int_{[-\pi, \pi)} \psi(T) \, d\mathcal{E}(e^{i\lambda}, e^{i\theta}), \qquad (T \in A),$$

where $\mathcal{E}$ is a resolution of the identity on $\mathcal{B}(\mathbb{T}^2)$ defined by $\mathcal{E}(C) = E(C \cap K)$ for all $C \in \mathcal{B}(\mathbb{T}^2)$. Note that by Equation (A.1) above, we see that

$$U_1^m U_2^n = \int_{[-\pi, \pi)} \int_{[-\pi, \pi)} e^{im\lambda + in\theta} \, d\mathcal{E}(e^{i\lambda}, e^{i\theta}), \qquad m, n \in \mathbb{Z}.$$

Bibliography

[1] O. P. Agrawal, D. N. Clark and R. G. Douglas, *Invariant subspaces in the polydisk*, Pacific J. Math., **121**, 1986, 1–11.

[2] N. Bansal, G. G. Hamedani and H. Zhang, *Non-linear regression with multidimensional indices*, Statistics and Probability Letters, **45**, 1999, 175–186.

[3] S. Cambanis and A. G. Miamee, *On prediction of harmonizable stable processes*, The Indian Journal of Statistics, Series A, **51**, 1989, 269–294.

[4] R. Cheng, *On the structure of shift-invariant subspaces of $L^2(\mathbb{T}^2, \mu)$*, Rocky Mountain J. Math., **24**, 1994, no. 4, 1307–1317.

[5] L. Gawarecki, V. Mandrekar and P. Richard, *Proper moving average representations and outer functions in two variables. Dedicated to Professor Nicholas Vakhania on the occasion of his 70th birthday*, Georgian Math. J., **8**, 2001, no. 2, 275–281.

[6] P. Ghatage and V. Mandrekar, *On Beurling type invariant subspaces of $L^2(\mathbb{T}^2)$ and their equivalence*, J. Operator Theory, **20**, 1988, 83–89.

[7] Paul Halmos, *Introduction to Hilbert Space and the Theory of Spectral Multiplicity*, Chelsea Publishing, 1957 (reprint by the A.M.S., 2000).

[8] Paul R. Halmos, *Shifts on Hilbert spaces*, J. Reine Angew. Math., **208**, 1961, 102–112.

[9] Clyde D. Hardin, *On the Spectral Representation of Symmetric Stable Processes*, J. Multivariate Anal., **12**, 1982, 385–401.

[10] Henry Helson and David Lowdenslager, *Prediction Theory and Fourier Series in Several Variables*, Acta Math., **99**, 1958, 165–202.

[11] Henry Helson, *Lectures on Invariant Subspaces*, Academic Press, 1964.

[12] G. Herglotz, *Uber Potenzreihen mit positivem reellen Teil im Einheitskreis*. Ber. Verh. Kgl. Sächs. Ges. Wiss., Leipzig, Math.-Phys. Kl., **63**, 1911, 501.

[13] Kenneth Hoffman, *Banach Spaces of Analytic Functions*, Prentice Hall, Inc., 1962.

[14] Karel Horák and Vladimír Müller, *On the structure of commuting isometries*, Comment. Math. Univ. Carolin., **28**, 1987, no. 1, 165–171.

[15] Yuzo Hosoya, *Harmonizable Stable Processes*, Z. Wahrscheinlichkeitstheorie verv. Gebiete, **60**, 1982, 517–533.

[16] Keiji Izuchi and Yasuo Matsugu, *Outer functions and invariant subspaces on the torus*, Acta. Sci. Math. (Szeged), **59**, 1994, 429–440.

[17] R. C. James, *Orthogonality and linear functionals in a normed linear space*, Trans. Amer. Math. Soc., **61**, 1947, 265–292.

[18] G. Kallianpur and V. Mandrekar, *Nondeterministic random fields and Wold and Halmos decompositions for commuting isometries*, Prediction theory and harmonic analysis, 165–190, North-Holland, Amsterdam, 1983.

[19] A. N. Kolmogorov, *Stationary sequences in Hilbert's space*, Bolletin Moskovskogo Gosudarstvenogo Universiteta. Matematika **2**, 1941. 40pp.

[20] D. Kundu and A. Mitra, *Asymptotic properties of least squares estimates of 2-D exponential signals*, Multidimensional Systems and Signal Processing, **7**, 1996, 135–150.

[21] T. L. Lai and C. Z. Wei, *A law of the iterated logarithm for double array of independent random variables with applications to regression and time series models*, Ann. Probability, **10**, 1982, 320–335.

[22] A. Makagon and V. Mandrekar, *The spectral representation of stable processes: harmonizability and regularity*, Probab. Theory Related Fields, **85**, 1990, no. 1, 1–11.

[23] V. Mandrekar, *Second order processes*, R-M-406, Department of Statistics and Probability, 1986 (unpublished).

[24] V. Mandrekar, *The work of Wiener and Masani on prediction theory and harmonic analysis. Connected at infinity. II*, 173–184, Texts Read. Math., 67, Hindustan Book Agency, New Delhi, 2013.

[25] V. Mandrekar, *The Validity of Beurling Theorems in Polydiscs*, Proc. Amer. Math. Soc., **103**, no. 1, 1988, 145–148.

[26] P. Masani, *Shift invariant spaces and prediction theory*, Acta Math., **107**, 1962, 275–290.

[27] M. B. Priestley, *The analysis of stationary processes with mixed spectra-I.*, Journal of the Royal Statistical Society, Series B, **24**, 1962, 215–233.

[28] M. B. Priestley, *The analysis of stationary processes with mixed spectra-II.*, Journal of the Royal Statistical Society, Series B, **24**, 1962, 511–529.

[29] M. B. Priestley, *Detection of periodicities*, Applications of Time Series Analysis in Astronomy and Meteorology, Chapman & Hall, 1997, 65–88.

[30] James Radlow, *The Validity of Beurling Theorems in Polydiscs*, Proc. Amer. Math. Soc., **38**, no. 2, 1973, 293–297.

[31] C. R. Rao, L. Zhao and B. Zhou, *Maximum likelihood estimation of 2-D superimposed exponential signals*, IEEE Trans. Acoust., Speech, Signal Processing, 42, **7**, 1994, 1795–1802.

[32] D. A. Redett, *S-invariant subspaces of $L^p(\mathbb{T})$*, Proc. Amer. Math. Soc., **133**, no. 5, 2004, 1459–1461.

[33] D. A. Redett, *"Beurling type" subspaces of $L^p(\mathbb{T}^2)$ and $H^p(\mathbb{T}^2)$*, Proc. Amer. Math. Soc., **133**, no. 4, 2005, 1151–1156.

[34] Walter Rudin, *Fourier Analysis on Groups*, Interscience, 1962.

[35] Walter Rudin, *Function Theory in Polydiscs*, Benjamin, New York, 1979.

[36] Walter Rudin, *Real and Complex Analysis*, McGraw-Hill, 1987.

[37] Walter Rudin, *Functional Analysis*, McGraw-Hill, International Series in Pure and Applied Mathematics, 1991.

[38] Harold S. Shapiro, *Topics in Approximation Theory*, Springer-Verlag, Lecture Notes in Mathematics, 187, 1971.

[39] Marek Slociński, *On the Wold-type decomposition of a pair of commuting isometries*, Ann. Polon. Math., **37**, 1980, no. 3, 255–262.

[40] K. Urbanik, *Random measures and harmonizable sequences*, Studia Math., **31**, 1968, 61–88.

[41] P. Whittle, *The simultaneous estimation of time series harmonic components and covariance structure*, Trabafos Estadistica, **3**, 1952, 43–57.

[42] P. Whittle, *The statistical analysis of a Seiche record*, J. Marine Research, **13**, 1954, 76–100.

[43] N. Wiener and P. Masani, *The prediction theory of multivariate stochastic processes. I. The regularity condition*, Acta Math., **98**, 1957, 111–150.

[44] N. Wiener and P. Masani, *The prediction theory of multivariate stochastic processes. II. The linear predictor*, Acta Math., **99**, 1958, 93–137.

[45] Herman Wold, *A study in the Analysis of Stationary Time Series*, Almqvist & Wiksell, Stockholm, 1954.

[46] Hao Zhang and V. Mandrekar, *Estimation of Hidden Frequencies for 2D Stationary Processes*, Journal of Time Series Analysis, **22**, no. 5, 2001, 613–629.

Index

Printed and bound by CPI Group (UK) Ltd, Croydon, CR0 4YY

24/10/2024

01778279-0004